Complete Library Skills

Grades K–2

Published by Instructional Fair
an imprint of

Author: Linda Turrell
Editor: Cary Malaski

 Children's Publishing

Published by Instructional Fair
An imprint of McGraw-Hill Children's Publishing
Copyright © 2004 McGraw-Hill Children's Publishing

All Rights Reserved • Printed in the United States of America

Limited Reproduction Permission: Permission to duplicate these materials is limited to the person for whom they are purchased. Reproduction for an entire school or school district is unlawful and strictly prohibited.

*The Dewey Decimal Classification® system is a registered trademark of Forest Press, Inc.

Send all inquiries to:
McGraw-Hill Children's Publishing
3195 Wilson Drive NW
Grand Rapids, Michigan 49544

Complete Library Skills—grades K–2
ISBN: 0-7424-1952-5

1 2 3 4 5 6 7 8 9 MAZ 09 08 07 06 05 04
The McGraw-Hill Companies

Table of Contents

Kindergarten
- The Library 4
- A Trip to Your Library 5
- What Will I Find? 6
- Your Librarian 7
- How to Care for a Book 8–12
- Library Citizenship Bookmarks..... 13
- Library Citizenship 14
- Using a Shelf Marker 15
- Putting Books on a Shelf 16
- Choosing a Book 17
- Returning a Book Rewards 18
- An Author's Job 19
- All in a Day's Work 20
- Story Time 21
- Do Your ABCs 22
- Alphabetizing by Author's Last Name 23
- Missing Books 24
- ABC Order 25
- Letter to Parents 26

First Grade
- Letter to Parents 27
- I Can Learn So Much! 28
- What Can I Do In a Library? 29
- A Helping Hand 30
- Library Citizenship 31
- Keys to Reading Success 32
- Library Citizenship Bookmarks..... 33
- Library Awards/Reminders 34
- Putting Books on a Shelf 35
- Neat Bookshelves 36
- Using a Shelf Marker 37
- What Is an Author? 38
- Finding the Author 39
- What Is an Illustrator? 40
- Find the Illustrator 41
- Teamwork 42
- Arranging Books in ABC Order 43
- Alphabet Fun 44
- ABC Dot-to-Dot 45
- In the Library 46
- ABCs to the Rescue 47
- The Alphabet Divided 48
- The Cover of a Book 49
- The Spine of a Book 50
- The Title Page 51
- The Table of Contents 52–55

- The Title of a Book 56
- Finding the Right Book 57
- Fiction and Nonfiction Books 58
- Fiction or Nonfiction? 59–61
- Call Numbers 62
- Using a Dictionary 63–64
- The Parts of a Computer 65
- Make an Alphabet Book 66
- Solve the Mystery 67
- Computer Reward 68

Second Grade
- A Place of Wonder 69
- Blast Off! 70
- Bookmarks 71
- Nine Tips for Choosing a Book ... 72
- The Title and Copyright Pages .. 73
- The Table of Contents 74
- The Parts of a Book 75
- Which Book Comes Next? 76
- Wait a Second! 77
- A Day at the Beach 78
- Jumpin' Jellyfish 79
- Who's Next? 80
- Nonfiction Books 81
- Fiction Books 82
- Biographies and Autobiographies 83
- Folktales and Fairytales 84
- Poetry 85
- Fiction vs. Nonfiction 86
- An Author's Name 87
- Fiction Call Numbers 88
- Arrange the Fiction Books 89
- Shelving Fiction Books 90
- Nonfiction Call Numbers 91
- Shelving Nonfiction Books 92
- Arranging Nonfiction Books 93
- My Spine 94
- The Ten Sections of the Library .. 95
- The Dewey Decimal Classification® System 96
- Which Section Is It? 97
- Dewey Match Game 98
- Dripping Dewey 99
- The Dictionary 100
- Using a Dictionary 101
- A-Maze-ing Spelling 102
- Which Is Right? 103
- The Encyclopedia 104

- Using an Encyclopedia 105
- Dictionary and Encyclopedia Review 106
- Guide Words 107
- Using Guide Words 108
- Guide Me 109
- Find the Guide Words 110
- Finding a Word 111
- Which Section Is It In? 112
- Dictionary Dividers 113
- Divide the Dictionary 114
- Round 'Em Up! 115
- Lasso Lily 116
- I'm Divided 117
- Know Your Computer 118
- Keyboard Practice 119
- You're an Artist 120
- A Nonfiction Book Project 121
- My Animal Report Notes 122
- How Did I Do? 123

- Library Awards/Reminders 124
- Newbery and Caldecott Book Award Winners 125
- Suggested Authors for K–2 126
- Answer Key 127–128

 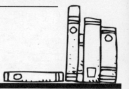

The Library

The library is a magical place
Where sometimes it feels like a race
To find the perfect book to call your own
To find the perfect book that you can take home.

But there are plenty of books to share with your friends
And don't worry if when library ends
You haven't found that perfect book
Because next week you can take another look.

Meanwhile, enjoy the book that you have selected
And make sure that you keep it well protected
From rain, tears, pets, and spills
And promise that when it's due you will
Return your book to the library right away
So your friends can enjoy it another day.

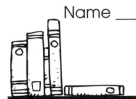

Name _____ Date _____

A Trip to Your Library

The library is full of things to explore. Just open the door and you can go anywhere!

The library is filled with all different kinds of books. Some have big, colorful pictures. Others have smaller pictures and a lot of words. There are tables and chairs and beautiful pictures on the walls. There might even be computers for you to use with an adult.

The librarian will read her favorite book to you and your friends. Then she will show you how to draw a clown, make a puppet, or hop like a kangaroo. Later you can look at all the exciting books. Take your time and pick out that special book just for you. The librarian will check it out for you. You can take this book home and look at it over and over until you go to the library again. Be sure to take good care of the book so other students can enjoy it when you bring it back.

Whenever you visit the library, there will always be a new book and a new adventure waiting for you. Come on in!

Name _____ Date _____

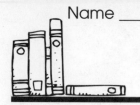 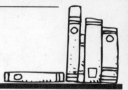

What Will I Find?

You will find many things in the library. Just look below.

▶ **Find the bookshelf. Color it red.**
Find the table. Color it blue.
Find the computer. Color it green.
Find the librarian. Color her yellow.

© McGraw-Hill Children's Publishing   0-7424-1952-5 *Complete Library Skills*

Name _____ Date _____

Your Librarian

I will be there as you start to explore
The wonders behind the library door.

I will help you find a book you will like
About a dog, a cat, or a mouse on a bike.

I will read a story to you and your friend.
We will then make puppets so we
can pretend.

I will show you all there is to know,
Then check out your book when it's time
to go.

I will wave and smile as you go out the door,
Knowing that soon you will be back for more.

I will hope that you treat your book with
great care,
And when you return it all the pages
are there.

I will always be here for your library needs
To help you discover more books to read.

Name _____ Date _____

How to Care for a Book

Taking Care of a Book

My "Read-at-Home" Book

Colored by _____

The first thing I do before reaching for a book is **wash my hands**. I want to make sure my hands are nice and clean before touching a book. Dirt belongs outside and food belongs in my stomach—not on a book.

1

Name _____ Date _____

 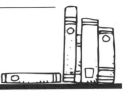

How to Care for a Book

The next thing I do is make sure my **crayons, markers, scissors, and glue are all far away from my book**. The pages already have pictures. I don't need to decorate them.

2

Sometimes I don't have time to finish my book. So I **mark my spot in the book with a bookmark**. Sometimes I use one of my favorite pictures as a bookmark. Other times I use a bookmark that my teacher gave me. But one thing I never do is bend a page to mark my spot.

3

How to Care for a Book

Turning a page too fast or too hard can sometimes rip the page. I always make sure **I turn the page slowly from the top corner**. When I do this, the page turns easily and will not rip.

4

I love to bring my library books home to show off. When I carry them home, **I always put them safely into a book bag or backpack**. That way, they won't get rained on and I won't accidentally drop them on the ground.

5

How to Care for a Book

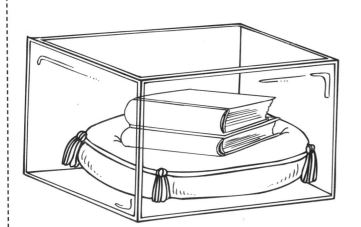

Once my books make it home safely, I take them out of my bag and I **put them in a safe place in my house**. My little brother doesn't get to play with them and my dog can't reach them. Since I take such good care of them at home, I will be able to return them to the library just like I found them.

I always remember to **return my book to the library on time**. Other kids might want to read it, so I need to finish it and bring it back when I'm supposed to. I always look forward to choosing a new book, too.

How to Care for a Book

When I check out a book, I take care of it by doing these things:

1. I wash my hands so they are nice and clean before touching my book.
2. I keep my school supplies away from my book.
3. I mark my place in a book with a bookmark. I don't bend the pages.
4. I turn the pages by using the top corners so they don't rip.
5. I always carry my book in a book bag or backpack.
6. I keep my book in a safe place at home.
7. I always return my books to the library on time.

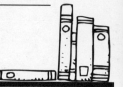

8

The End.

9

Name _____ Date _____

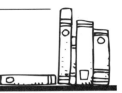

Library Citizenship Bookmarks

▶ **Reproduce the bookmarks and laminate them for long-term use.**

Remember to use an *indoor voice* when you are in the library.

Books can take you *anywhere!*

_____ was a good listener during *story time!*

Name _____ Date _____

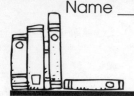
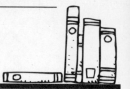

Library Citizenship

Cara and Nathan are twins. They are very alike. But when they go to the library, something happens and they act very different. Cara is quiet and Nathan is very noisy.

▶ **Look at the behaviors below. Cut each one out and paste it underneath the twin that acts that way. Then answer the question.**

Quiet Cara   Noisy Nathan

▶ **Which twin is showing good library behavior?** Cara   Nathan

talking	raising hand
not talking	talking out of turn
moving all over	throwing a book
sitting still	gently holding a book

© McGraw-Hill Children's Publishing   14   0-7424-1952-5 Complete Library Skills

Using a Shelf Marker

Part of using a library is taking books off of shelves to look at them. But it is important for your students to know that all books have a home—just like they do. That is why books must be put back exactly where they came from—in their homes. This helps keep the library organized and neat.

Show your students how to use a shelf marker to mark a book's place on the shelf. Ask your students what might happen if they put a book in another book's home. Would someone else be able to find it if it's not in its own home?

Reproduce this shelf marker on heavy construction paper or have your students color it and then laminate it for long-term use.

Name _____ Date _____

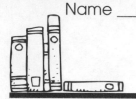
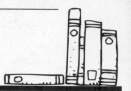

Putting Books on a Shelf

▶ **Look at the bookshelves below. Some of the books were put on the shelves neatly. Other books were put on the shelves incorrectly. Color the books that have been put on the shelves the right way.**

© McGraw-Hill Children's Publishing — 0-7424-1952-5 Complete Library Skills

Name _____ Date _____

Choosing a Book

When you go to the library, you will see a lot of books. How will you know which ones to pick up and which ones to pass by? Look at the cover. If it looks interesting, pick it up.

➡ **Look at the book covers below. Color the books that you would pick up.**

Name _____ Date _____

Returning a Book Rewards

Name _____ Date _____

An Author's Job

An author tells a story with words.

▶ **Look at the pictures below. Write words to go with each picture.**

19

© McGraw-Hill Children's Publishing

0-7424-1952-5 *Complete Library Skills*

Name _____ Date _____

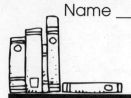 

All in a Day's Work

An illustrator draws pictures to go with a story.

▶ **Look at the words below. Draw pictures to go with the words.**

Kim and Ben go to the park.

Kim plays on the swing.

Ben likes the slide.

Kim and Ben go home happy.

Story Time

One of the most special times of the day for a young child is story time. During this time, you can create an educational experience that is a lot of fun and one that will motivate children to want to explore books.

The Benefits of Story Time

1. We all hope that children will enter school with a beginning knowledge and appreciation of books, but unfortunately this is not always true. There are many young children who start school and have not been afforded the opportunity of having books in their environment. The school library and the stories told at school may be a child's first introduction to the world of books.
2. The story hour and the books you choose to read or "tell" help to teach an appreciation of literature.
3. Storytelling is one of the most effective means of increasing language development in young children. Not only are listening skills increased, but expressive vocabulary begins to soar when storytelling is incorporated into a child's daily life.
4. Story time is an excellent tool for increasing listening or "paying attention" skills.
5. Cultural traditions of people can be taught through storytelling.
6. Storytelling can also be used to teach and reinforce academic subjects such as reading, math, science, social studies, etc.
7. And, of course, storytelling will entertain, amuse, and delight young children!

How to Prepare for Story Time

1. Read the story through once, just for your own enjoyment.
2. Read it again for the sequence of events and pay special attention to the outline of the plot.
3. Find some key phrases, such as "Fee-fi-fo-fum," or "I'll huff and puff and blow your house down." When key phrases are used in the story, ask the children to say those key phrases with you.
4. Practice telling the story. Read it over several times.
5. The ending is very important. Strive to retain the mood of your story. If it is exciting use a lifting voice. If it is serious, use a sober tone.
6. Help the children visualize the story as you go along, and bring the characters alive for them.

Some Don'ts During Story Time

Do not speak too fast and don't be too dramatic. Don't explain in too much detail, or you might lose your audience. And don't make story time longer than their attention spans can handle. Even if they ask for more, leave them wanting more.

Name _____ Date _____

Do Your ABCs

Many books are kept in alphabetical order in the library. This is done by the first letter of the author's last name. Do you know the alphabet?

▶ **Connect the dots starting with A.**

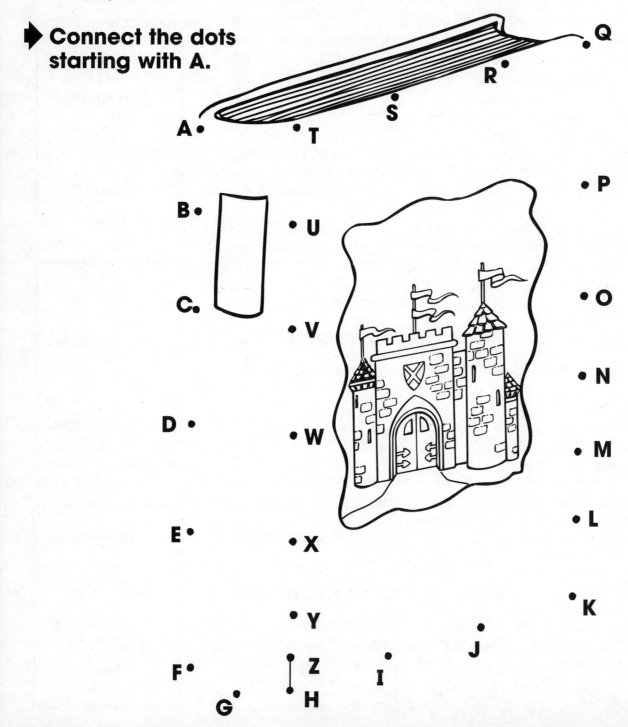

Name _____ Date _____

Alphabetizing by Author's Last Name

▶ **The bookshelves are empty. Cut out the books below. Glue them in alphabetical order using the author's last name.**

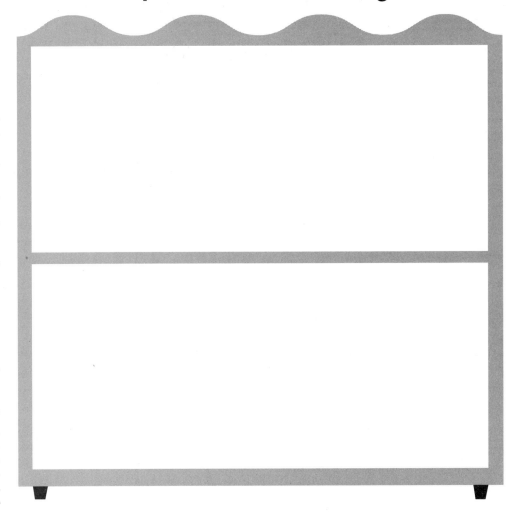

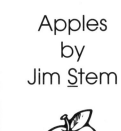
Apples
by
Jim <u>S</u>tem

Ice
by
Amy <u>F</u>reeze

Trot Along
by
Neil <u>N</u>ay

Sticky Stuff
by
Johnny <u>J</u>elly

Swish Swish
by
Tim <u>B</u>ubble

Puppies
by
Judy <u>R</u>uff

Name _____ Date _____

Missing Books

▶ **Look at the books below. Each book has the first letter of the author's last name on it. Some books are missing. Write in the missing letters in alphabetical order.**

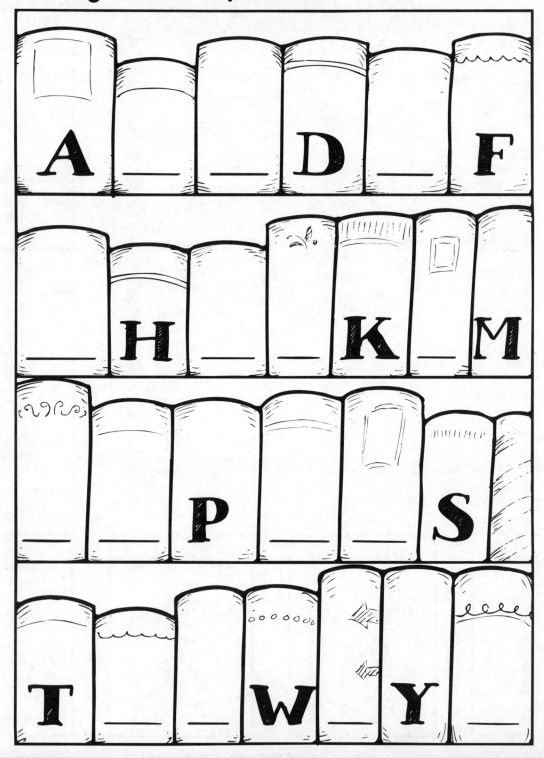

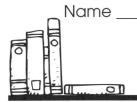 

ABC Order

▶ **The library has many books. The books are placed on the shelves in ABC order. This is done by the author's last name. Can you connect a line from each book to its shelf?**

 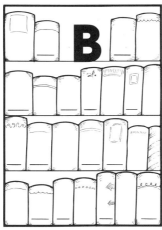 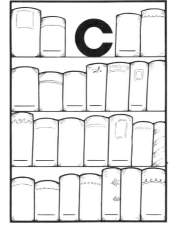

Dear Parents,

Today your kindergartener brought home a book from our school library for the first time. This is an exciting time for a young child. Being able to choose a book, take it home, and care for it is one of the first "grown-up" responsibilities a child is given.

We have spent time discussing proper library behavior and talking about how to carefully handle a book. However, it is important that you reinforce and model these "grown-up" behaviors.

Mark on your calendar (at your child's eye level, if possible) your child's assigned library day. This day is _____. This is the day your child should bring to school any books he or she has checked out from the school library.

Please make sure that your child carries his or her books to school in a book bag or a backpack. Tucking books into some sort of bag makes carrying them easier for a small child and it will help protect the books from unnecessary wear or damage, especially in inclement weather.

Learning to read is an exciting experience. You can foster this attitude by reading to your child whenever possible and by encouraging your child to pick up books and explore them. This special time can be a memorable experience for both you and your child.

Thank you for your support,

Dear Parents,

The most exciting part of first grade is learning to read. It is a magical experience for a young child. Your child will want to show off this new skill when he or she comes home.

When your child brings home a book from the library, or reads a book that you may have at home, sit with your child and be his or her audience. Help your child with proper names, uncommon words, or words that may be difficult to sound out.

Discuss the story as you are reading or when you are finished. "What part of the story was your favorite?" "Who was your favorite character?" "Did you like the way the story ended?" "Can you think of another ending you would like better?" There are so many questions you can ask your child that will help him or her recall the story and use his or her imagination. Questioning your child at the end of a story will help him or her grasp a better understanding of the story.

In first grade, we expect children to begin taking more responsibility for caring for their library books, and in making their own book selections. Your child's library day is _____. Although children at this age want to be more independent, they will still need your reminders about returning library books on time. Post library day on your refrigerator and encourage your child to try and remember when to return books.

I look forward to an exciting year with your "eager-to-read" first graders.

Thank you for your support,

Name _____ Date _____

I Can Learn So Much!

A library is filled with books that are fun to read. But you can also learn about almost anything in a library.

▶ **Cut out the boxes below. If you could learn about it in a library, glue it in the book. If you could not learn about it in a library, glue it outside of the book.**

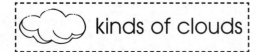 kinds of clouds

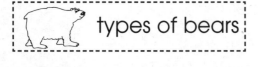 types of bears

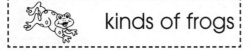 kinds of frogs

 your phone number

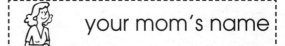 your mom's name

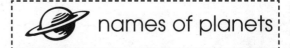 names of planets

Name _____ Date _____

What Can I Do In a Library?

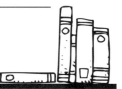

You can use a library for many things.

▶ **Look at each picture below. If it shows something you would do in a library, color the 🚦 green. If it shows something you would not do in a library, color the 🚦 red.**

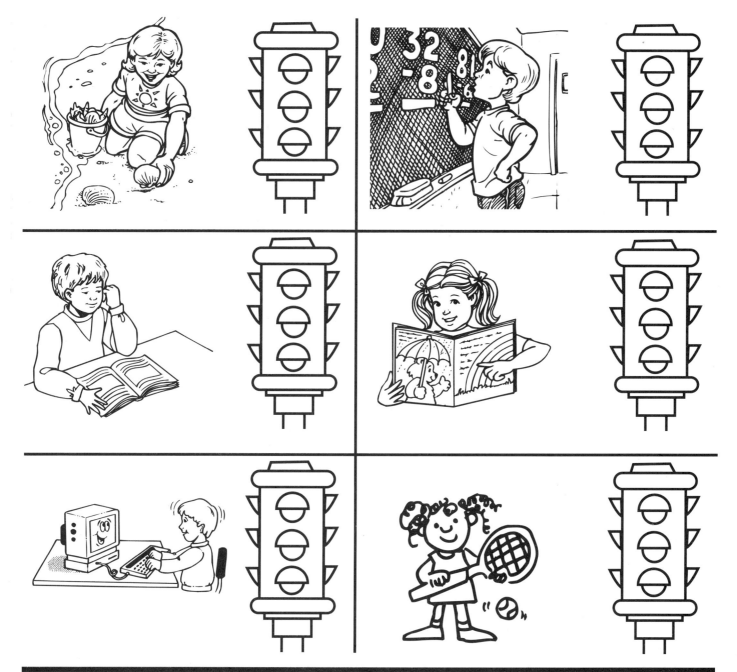

© McGraw-Hill Children's Publishing   0-7424-1952-5 Complete Library Skills

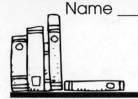
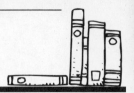

A Helping Hand

A librarian does many things in the library.

▶ **Look at each picture below. If you would see a librarian doing it in a library, circle it. If you would not see a librarian doing it in a library, put an X across it.**

Name _____ Date _____

Library Citizenship

When you come to the library, it is important that you bring your library manners with you. Do you know your library manners?

➤ **Look at the pictures below. If the student is using library manners, color the picture.**

Name _____ Date _____

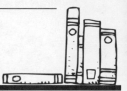

Keys to Reading Success

▶ **Read the book behaviors below. If it is a good book behavior, color the space red. If it is a bad book behavior, color the space black.**

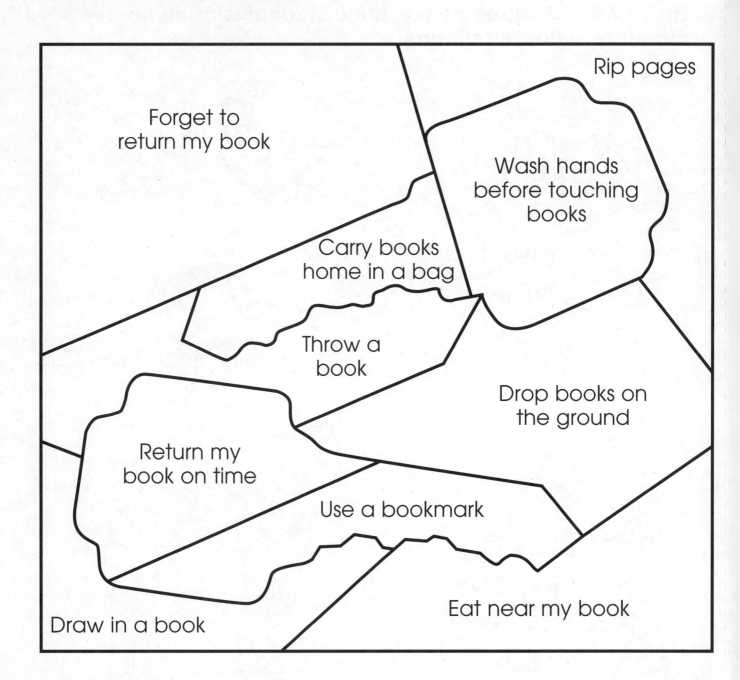

- Rip pages
- Forget to return my book
- Wash hands before touching books
- Carry books home in a bag
- Throw a book
- Drop books on the ground
- Return my book on time
- Use a bookmark
- Draw in a book
- Eat near my book

© McGraw-Hill Children's Publishing

0-7424-1952-5 Complete Library Skills

Name _____ Date _____

Library Citizenship Bookmarks

- Keep your books **safe** from bad weather.
- **Carry them in a book bag.**
- keep your books clean. Do not write, scribble, or color in them.
- Keep food **AWAY** from your books.
- Return your books to the library on time.

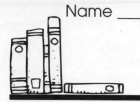 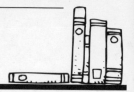

Library Awards/Reminders

▶ **Students love earning awards. Use the awards below to reward a student for returning his or her book on time. Or use the reminder to let students know that their library books will be due soon.**

Name _____ Date _____

Putting Books on a Shelf

When you pull a book from a shelf in the library, it is important to put it back correctly. Books should stand straight and tall on the shelves.

➡ **Look at the sentences. Use the Word Bank to fill in each missing word. Write the word on the blank. Each sentence tells how to correctly put books on a shelf.**

Word Bank
book
marker
tall
out
gently

1. Find where you put your shelf _____.

2. Pull _____ your shelf marker.

3. Slide your book _____ onto the shelf.

4. Is your _____ facing the right way?

5. Make sure it stands nice and _____.

Name _____ Date _____

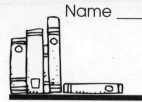
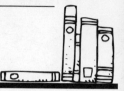

Neat Bookshelves

Every book in the library has its very own place on the bookshelves. It is important to always put it back so it faces the right way and stands straight up.

▶ **Look at the full bookshelves on the left. Color the books that were put on the shelves incorrectly. Then draw books on the empty bookshelves on the right. Make sure they stand straight and face the right way.**

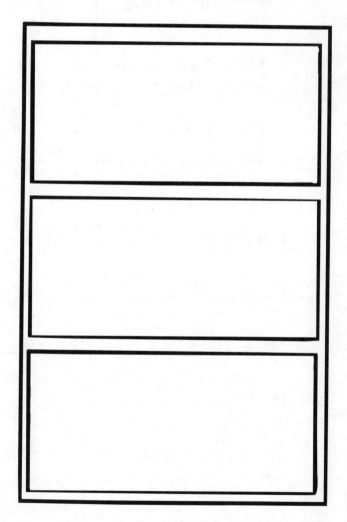

© McGraw-Hill Children's Publishing

0-7424-1952-5 Complete Library Skills

Name _____ Date _____

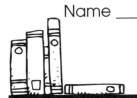

Using a Shelf Marker

▶ **Look at the bookshelves below. A book has been removed from each shelf. Cut out the shelf markers and glue them where they should be placed.**

 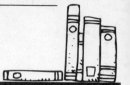

What Is an Author?

An author is someone who writes a book. You can find his or her name on the cover of the book, on the spine of the book, or on the inside title page.

▶ **Look at the pictures. Under each picture, write what you think is happening. Then cut around the outside of the page and fold it. Now you are an author, too.**

Name _____ Date _____

Finding the Author

▶ **Look at the books. Circle the author's name on each one.**

The Milky Way Galaxy by Star Myers

On the Move by Jerry Geranium

Life in a Castle by Shiny Armor

▶ **Look around your classroom or library. Find a book you like. Write the title and author on the lines below.**

_____
Title

_____
Author

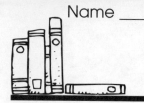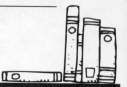

What Is an Illustrator?

An illustrator is someone who creates the pictures in books and on the covers. His or her name can be found on the cover of the book or on the inside title page. Look around your classroom or library at books. Notice how different they all look.

▶ **Read the sentences below. Draw pictures to go with each one. Cut the page out and fold it. Now you are an illustrator, too.**

Lillie, the library lizard, likes to read books.

Lillie has many books.

Name _____   Date _____

Find the Illustrator

You can tell a lot about a book from its illustrations on the cover.

▶ **Look at the books below. They are missing their titles. Find the correct title on the right and draw a line to connect it to its book.**

Animal Homes
by
Jack Arbor

The Storm of the Century
by
Ima Blizzard

On the Dock
by
I.M. Relaxing

My Dog Doesn't Walk
by
U.R. Spoiled

Name _____ Date _____

Teamwork

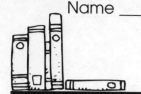

An author and an illustrator need to work well together when making a book.

▶ **With a partner, write and illustrate the book below. One of you can write and one of you can illustrate. Cut it out and staple the pages together when you are done.**

Written by:

Illustrated by:

Name _____ Date _____

Arranging Books in ABC Order

Some books are arranged in the library in ABC order by the author's last name.

▶ **Connect the dots below in ABC order.**

Alphabet Fun

➤ **Fill in the missing letters below. Then use the code to find your answer to this riddle:**

What has two legs and one head and follows you everywhere?

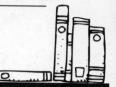

$\dfrac{a}{4} \quad \dfrac{b}{13} \quad \dfrac{\_}{6} \quad \dfrac{\_}{1} \quad \dfrac{e}{10}$

$\dfrac{\_}{16} \quad \dfrac{g}{3} \quad \dfrac{\_}{25} \quad \dfrac{\_}{9} \quad \dfrac{\_}{14}$

$\dfrac{\_}{8} \quad \dfrac{\_}{21} \quad \dfrac{\_}{19} \quad \dfrac{n}{12} \quad \dfrac{\_}{26}$

$\dfrac{p}{11} \quad \dfrac{\_}{23} \quad \dfrac{\_}{2} \quad \dfrac{\_}{20} \quad \dfrac{t}{17}$

$\dfrac{\_}{24} \quad \dfrac{\_}{7} \quad \dfrac{w}{15} \quad \dfrac{\_}{5} \quad \dfrac{\_}{22} \quad \dfrac{\_}{18}$

Answer to the riddle:

$\dfrac{\_}{22} \; \dfrac{\_}{26} \; \dfrac{\_}{24} \; \dfrac{\_}{2} \quad \dfrac{\_}{20} \; \dfrac{\_}{25} \; \dfrac{\_}{4} \; \dfrac{\_}{1} \; \dfrac{\_}{26} \; \dfrac{\_}{15}$

© McGraw-Hill Children's Publishing 44 0-7424-1952-5 Complete Library Skills

Name _____ Date _____

ABC Dot-to-Dot

➡ **Connect the dots in alphabetical order. Then color the picture.**

little

kite

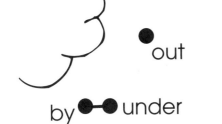
out

by • — • under

pan

swim it hay

dog eye

Name _____ Date _____

In the Library

➡ **Cut out the words on the right. Glue them in alphabetical order on the lines.**

Alphabetical List

1. _____
2. _____
3. _____
4. _____
5. _____
6. _____
7. _____
8. _____

illustrator
stories
author
librarian
read
books
quiet
computer

© McGraw-Hill Children's Publishing   0-7424-1952-5 Complete Library Skills

Name _____ Date _____

ABCs to the Rescue

▶ **Max lost his puppy. He needs your help finding it. Trace a path to the puppy in alphabetical order.**

▶ **Where did Max find his puppy? Circle the word that comes last alphabetically in each box. Read those words together to tell you where he found his puppy.**

big	in	after	pond
more	an	the	lake
swimming	digging	now	jump

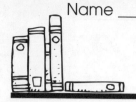
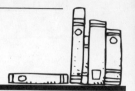

Name _____ Date _____

The Alphabet Divided

The alphabet can be divided into three sections. The beginning is letters A–H. The middle is letters I–Q. And the end is letters R–Z.

▶ **Read the words in the Word Bank. Look at the first letter of each word. Write them in the correct sections below.**

Word Bank

jump	food	keep
ate	help	look
ripe	say	vase

Beginning A–H	Middle I–Q	End R–Z

Name _____   Date _____

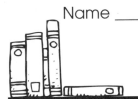 

The Cover of a Book

A book cover is the first thing you see when you look at a book. It can tell you a lot about a book. But it also keeps the inside pages safe from tearing or getting dirty.

▶ **Draw the cover of your favorite book below.**

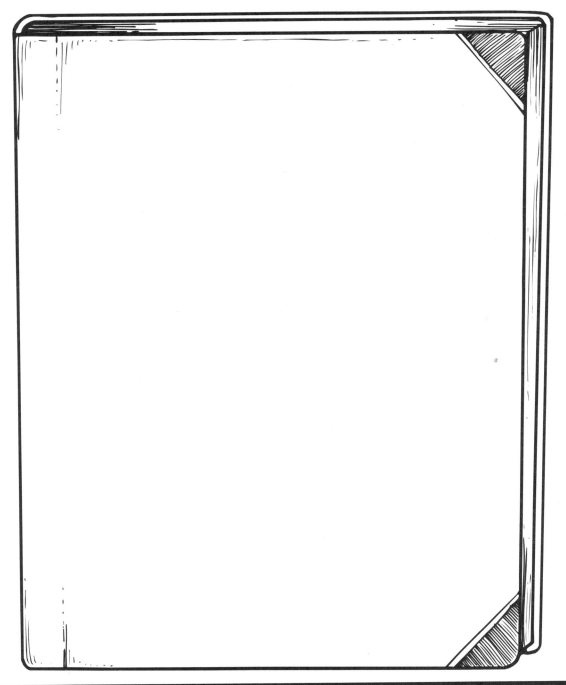

Name _____ Date _____

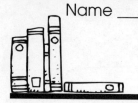
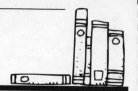

The Spine of a Book

All books have a spine. It helps hold the pages of a book together.

▶ **Write the title of your favorite book on the spine below. Then color the book.**

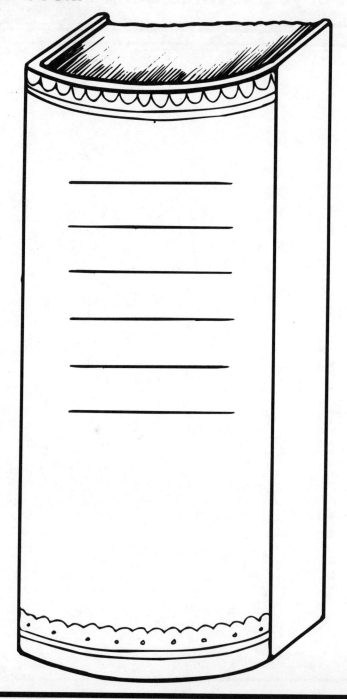

© McGraw-Hill Children's Publishing 0-7424-1952-5 Complete Library Skills

Name _____ Date _____

The Title Page

One of the first pages inside a book is called the title page. The title page tells us the title of the book and the author.

▶ **Read the title page below. Answer the questions.**

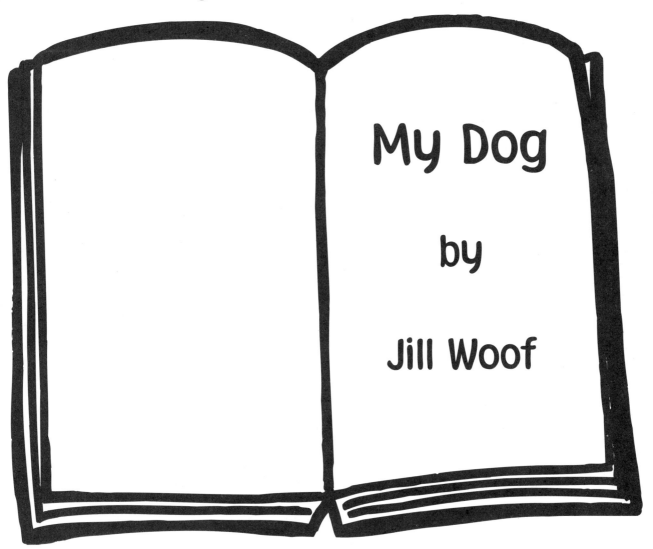

What is the title of this book? _____

Who is the author of this book? _____

Name _____ Date _____

The Table of Contents

Another page at the beginning of a book is called the table of contents page. This page helps you finds things in a book.

▶ **Read the table of contents below. Answer the questions.**

Table of Contents

Bats .... page 3

Cats .... page 5

Bees .... page 7

On what page will you find cats? _____

On what page will you find bats? _____

On what page will you find bees? _____

Name _____ Date _____

The Table of Contents

➤ **Read the table of contents below. It will tell you where to find things in a book. Then answer the questions.**

Table of Contents

Frogs............................page 3

Bugs ....page 5

Fish ....page 7

On what page will you find fish? _____

On what page will you find frogs? _____

On what page will you find bugs? _____

Name _____ Date _____

The Table of Contents

▶ **Read the table of contents below. It will tell you where to find things in a book. Then answer the questions.**

Table of Contents

Stars .... page 3

Comets .... page 6

The Moon .... page 9

The Sun .... page 12

The Planets .... page 15

On what page will you find comets? _____

On what page will you find the moon? _____

On what page will you find the sun? _____

What will you find on page 3? _____

What will you find on page 15? _____

Name _____   Date _____

The Table of Contents

▶ **Read the table of contents below. It will tell you where to find things in a book. Then answer the questions.**

Table of Contents

Sharks .... page 3

Whales .... page 6

Octopuses .... page 9

Seals .... page 12

Penguins .... page 15

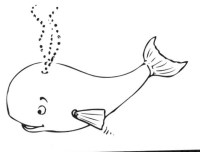

What will you find on page 6? _____

What will you find on page 12? _____

On what page will you find sharks? _____

On what page will you find penguins? _____

On what page will you find octopuses? _____

The Title of a Book

Young students need to be taught how to look for books. Take some time and discuss book titles with your first graders.

Guess What the Book Is About

Show your students a book and read its title to the class. Ask your students what they think the book will be about. Repeat this with several books.

Making Up Book Titles

Another fun activity is to give your students a subject and have them think of titles for that subject. For example: If you gave your students the subject *trains*, they might come up with "Everything I Want to Know About Trains," or "The Train Who Was Afraid to Chug."

Sorting Books by Title

Fill a large tabletop with a variety of books. The books should cover two or three different subjects, such as pets, weather, or airplanes. Have your students sort the books by subject.

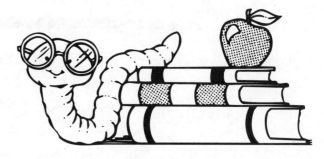

This is the first step in learning how to use the library as a research or reference tool. Young students will begin to learn that they can find information in the library.

Finding the Right Book

▶ You need a book. Look at the kind of book you need. Then look at your choices. Color the right book.

Cats

Bees

Pins

  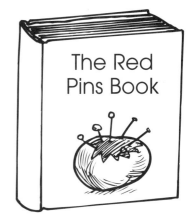

Name _____ Date _____

Fiction and Nonfiction Books

Your school library has many kinds of books. One kind of book is a storybook. Storybooks often have make-believe stories. Make-believe stories, or stories that are about things that are not real, are called **fiction** books. Books that tell you about real things, or facts, are called **nonfiction** books.

▶ **Look at the books below. Which one is a fiction book? Which one is a nonfiction book? How can you tell? Circle the fiction book. Draw a square around the nonfiction book.**

Taking Your Puppy to the Vet tells about something real. This is the nonfiction book.

The Mouse and the Talking Cheese is not real. Cheese cannot talk. This story is fiction.

Name _____ Date _____

Fiction or Nonfiction?

▶ **Read the book titles below. Color the fiction books.**

Name _____ Date _____

Fiction or Nonfiction?

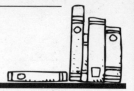

▶ **Read the book titles below. Color the fiction books.**

Name _____ Date _____

Fiction or Nonfiction?

▶ **Read the book titles below. Color the nonfiction books.**

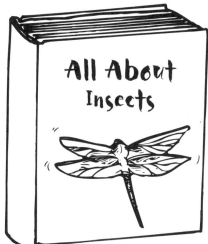

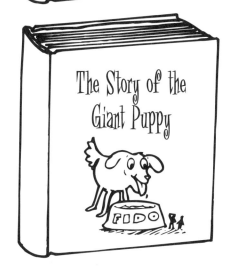

 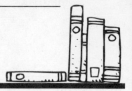

Name _____ Date _____

Call Numbers

On the spine of a fiction book, you will see two letters. These are called **call numbers**. They are the first two letters of the author's last name. The books are put on the shelves in alphabetical order using the call numbers.

If you saw a book by Peter Brook, you would see the letters **Br** on the spine. **Br** are the first two letters of the author's last name.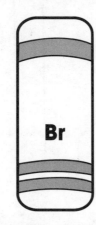

If you saw a book by Jane Page, you would see the letters **Pa** on the spine. **Pa** are the first two letters of the author's last name.

▶ **Write the call numbers for the books below.**

1. *The Dog* by Jim Allen

2. *My Teddy Bear* by Susie Flop

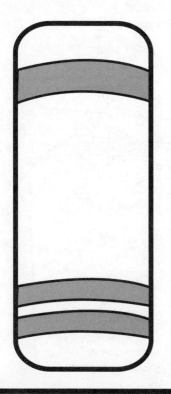 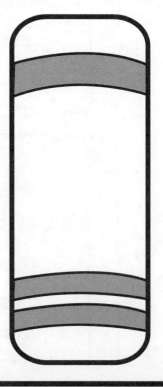

Name _____ Date _____

Using a Dictionary

A **dictionary** is a book of words. It tells you what words mean and helps you check how to spell them. The words in a dictionary are in alphabetical order.

To find a word, look at the first letter of the word. Is it near the front of the alphabet? Is it in the middle? Or is it near the end? That will tell you where to look in the dictionary.

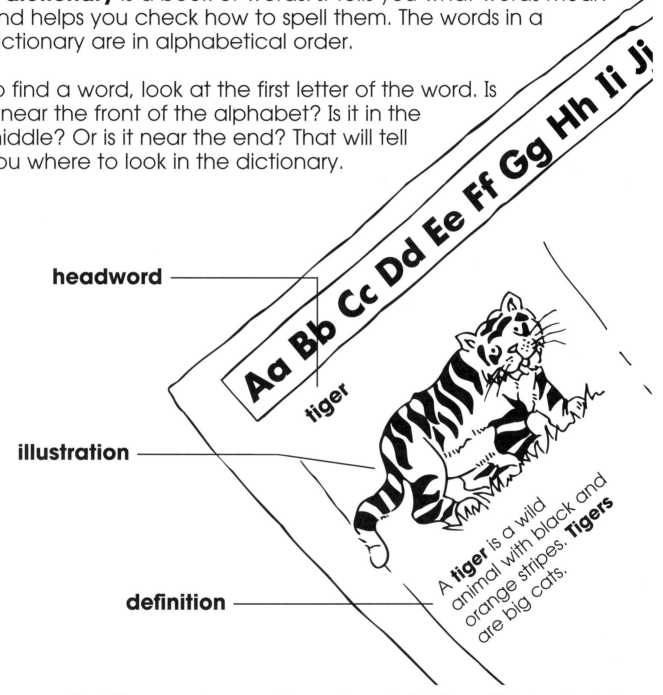

Once you find the word, you will see the **definition**. This is what the word means. Some words are illustrated with pictures that help you understand the word, too.

Name _____ Date _____

Using a Dictionary

Words in a dictionary are in alphabetical order. Each word has a **definition**. This tells you what the word means. Many words also come with a picture.

▶ **Look at the pictures below. Cut them out. Glue them under the matching letter.**

© McGraw-Hill Children's Publishing

Name _____ Date _____

The Parts of a Computer

You might find your computer lab in your school library. You can do many things on a computer. You can draw, color, type, and find information on a computer. But first you have to know the parts of a computer.

Monitor—This is the screen that shows you what you have done. It looks like a small TV. Color the monitor red.

Keyboard—This is where you will find the keys with the letters and numbers. Color the keyboard yellow.

Mouse—This is what you move around on a table. You use it to point to things on the screen. Color the mouse blue.

CPU—This is the "brains" of the computer. It does what you tell it to do. Color the CPU black.

Printer—This machine copies what is on the screen onto a piece of paper. Color the printer green.

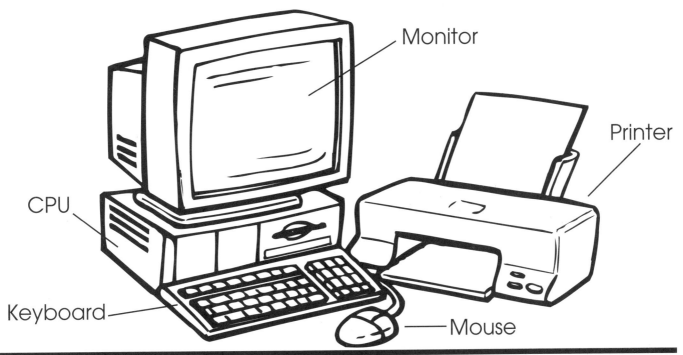

© McGraw-Hill Children's Publishing　　0-7424-1952-5 Complete Library Skills

Make an Alphabet Book

Use the following activity with your students after they have been introduced to your school's computer drawing program.

Your students can create a classroom alphabet book using the letters and stamps in your school's drawing program. This activity addresses the following skills: letter recognition, recognizing initial sounds, and following directions.

1. To prepare for the activity, read several alphabet books to your students. Tell students they will create their own alphabet books using the computer.
2. Assign each student a letter of the alphabet. Hand out copies of capital letters, if needed.
3. When your students sit down at their computers, have them open the drawing program used by your school.
4. Have them select the alphabet text tool [A] and type the letter they have been assigned.
5. Also have your students type their name at the bottom of the page.
6. Now have students stamp pictures onto the page that start with the same sound as the letter they were given.
7. Require your students to stamp ten stamps for their letter.
8. Have your students print their pages after you have checked their work.
9. Quit or exit the program.
10. Assemble all of the pages into a classroom alphabet book and display it in your classroom library.

Solve the Mystery

Use the following activity with your students after they have been introduced to your school's drawing and word processing programs.

Students love a good mystery, especially when they are the detectives. In this activity, students will type answers to a series of questions and then use the first letter of each answer to find a mystery word. This activity addresses the following skills: spelling, critical thinking, and following directions.

1. Have your students open the drawing or word processing program found on your school's computers. They will need to click on the text tool [A] so that they can stamp out words.

2. Either hand out the questions below so that your students can do this activity independently, or read aloud the following questions, giving your students time to type their answers:

   - What is the name of a creature that lives in the ocean and has eight arms?
   - What is the name of a sea creature that is small, flat, and has pinchers?
   - What is the name of a slithery creature that looks like a snake but lives in water?
   - What is the name of a sea creature that looks like a colorful flower?
   - What does a fisherman use to scoop fish out of the water after they are caught?

3. Now have them type the first letter of each answer, keeping them in the same order as above. This should spell out the answer to the following mystery:
   - What is wet and salty?

4. Make sure your students have typed their names on their pages.

5. Print the pages.

6. Quit or exit the program.

© McGraw-Hill Children's Publishing — 0-7424-1952-5 Complete Library Skills

Name _____   Date _____

Computer Reward

Congratulations!

_____

is a
Computer Superstar!

A Place of Wonder

Welcome to your school library, boys and girls. I am Miss Vespia and I am your librarian. I am so excited to have you visiting me today.

Did you know that the library is filled with many wonderful treasures and mysteries just waiting to be discovered? It's true! Dive into the deep blue ocean and take a journey into the unknown. Follow kids into a magical attic, and see what happens when they close the door. Imagine if your teacher had an identical twin, and you couldn't tell them apart. With books, dogs can talk, people can fly, and parents can see through walls. Books can make anything happen.

But these treasures and mysteries can only be uncovered if you look for them. When you come to the library, I want you to become like detectives. I want you to look for a book that will take you on a magical journey. I want you to go somewhere you've never been, do something you've never done, or learn about something new. That is the power of a book.

When you come to the library, I want you to bring your Book Detecto device. It will help you keep your place once you have found just the right book for you. If you take good care of it, it will help you as you take off on an unforgettable adventure.

Circle the best summary of the selection above.

a. The library is a place that is filled with boring, hard-to-find books and it would be a waste of my time to go.
b. The library is a confusing place and I won't find anything I like.
c. The library is filled with books that will take me on exciting adventures and teach me new things.

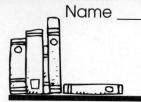

Name _____ Date _____

Blast Off!

▶ **If the space shows good library or book behavior, color it yellow. If the space shows things you should not do in a library, color it red.**

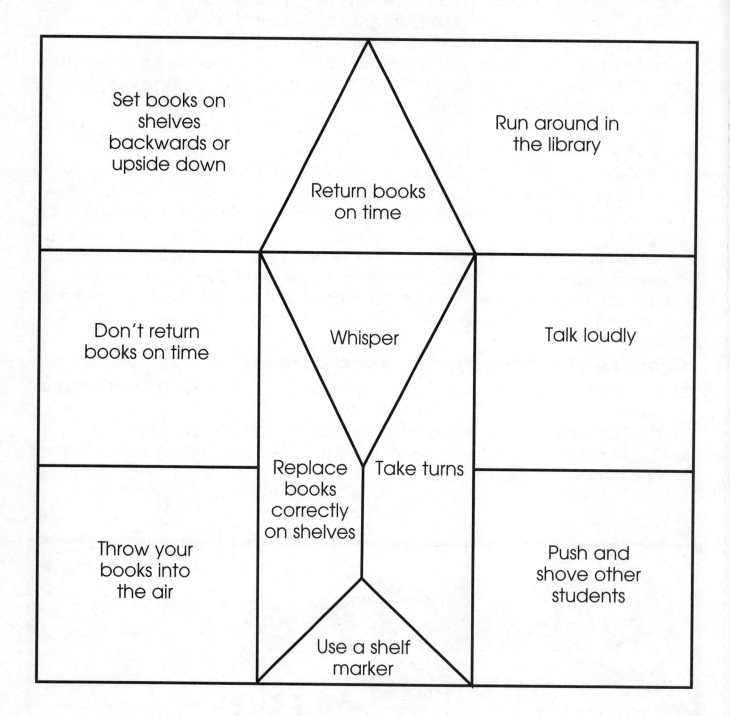

- Set books on shelves backwards or upside down
- Return books on time
- Run around in the library
- Don't return books on time
- Whisper
- Talk loudly
- Replace books correctly on shelves
- Take turns
- Throw your books into the air
- Push and shove other students
- Use a shelf marker

Name _____ Date _____

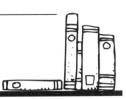

Bookmarks

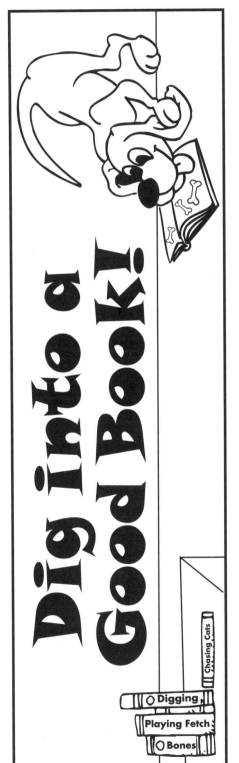

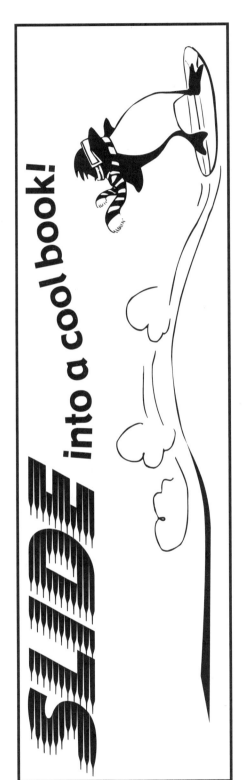

Dig into a Good Book!

Become a reader for life.

SLIDE into a cool book!

© McGraw-Hill Children's Publishing   71   0-7424-1952-5 *Complete Library Skills*

Name _____   Date _____

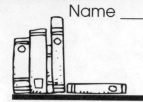
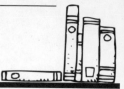

Nine Tips for Choosing a Book

Libraries have so many books! Sometimes it is hard to know where to start looking for a book. Read the tips below. They should help point you in the right direction.

1. Decide if you want your book to be fiction (make believe) or nonfiction (real).
2. Think of a subject that interests you or that you want to learn more about.
3. Look for books in a series that you really like.
4. Look for books by an author that you like.
5. Think of books that friends have told you about.
6. Find the courage to try something new.
7. Look at books on display in the library.
8. Ask your teacher for a recommendation.
9. Ask your librarian for a recommendation.

▶ **Take the Five Finger Test!**

Before you check out a book, take this test to make sure the book is right for you. Flip to a page in the book. For every word that is too difficult for you, put up a finger. By the end of the page, if you have all five fingers up, the book is too difficult for you. If you don't have any fingers up, the book is too easy for you.

Name _____ Date _____

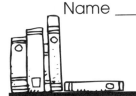 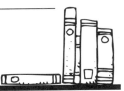

The Title and Copyright Pages

The title page and copyright page can tell you a lot about a book. They tell you the title, the author's name, where the book was made, and who made it.

➤ **Look at the title and copyright pages below. Answer the questions about them.**

1. What is the title of this book? _____
2. Who is the author? _____
3. Where was the book made? _____
4. Who made the book? _____
5. Can you read about cats in this book? _____
6. Can you read about trains in this book? _____
7. Can you read about puppies in this book? _____

Name _____ Date _____

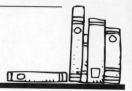

The Table of Contents

The table of contents page tells you where to find things in a book.

➤ **Read the table of contents below. Then answer the questions.**

Table of Contents

Chapter 1—Buzzing Bees .... page 3
Chapter 2—Small Seeds .... page 6
Chapter 3—Tall Trees .... page 9
Chapter 4—Vegetable Plants .... page 12
Chapter 5—Deep Roots .... page 15
Chapter 6—Strong Stems .... page 18
Chapter 7—How to Avoid Bees .... page 21
Chapter 8—How to Plant a Garden .... page 24
Chapter 9—When to Pick Vegetables .... page 27

1. On what page can you read about stems? _____
2. Where would you look to learn about vegetable plants? _____
3. What would you learn about on page 3? _____
4. What is the title of Chapter 9? _____
5. Chapter 2 starts on what page? _____
6. How many chapters are in this book? _____
7. Which chapter would you turn to if you wanted to learn about planting a garden? _____
8. About how many pages does this book have? _____
9. Is there a chapter about rain? _____
10. What might be a good title for this book? _____

Name _____ Date _____

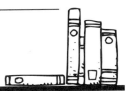

The Parts of a Book

➤ **Use the Word Bank to fill in the blanks.**

Word Bank

title page	author	cover
copyright page	illustrator	spine
publisher	table of contents	title

1. The name of a book is its _____.

2. The _____ tells you where to find things in a book.

3. The person who wrote the book is called the _____.

4. The _____ tells you when and where the book was made.

5. The part of a book that holds the pages together is called the _____.

6. The drawings and pictures in a book are done by the _____.

7. The _____ tells you the title and author of a book.

8. The _____ protects the pages inside a book.

9. The company that makes the book is called the _____.

Name _____ Date _____

Which Book Comes Next?

Books are put on bookshelves in alphabetical order by the author's last name. Sometimes there are last names that start with the same letter. When this happens, you have to look at the second letter in the author's last name to decide which one comes first.

▶ **Look at the bookshelves below. Cut out the book that is next on the shelf and glue it on.**

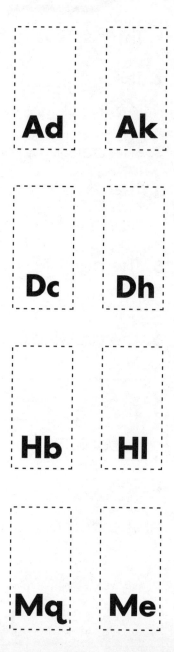

© McGraw-Hill Children's Publishing   0-7424-1952-5 Complete Library Skills

Name _____ Date _____

Wait a Second!

▶ **Each list below starts with the same letter. Circle the second letter in each word. Then number the words in alphabetical order.**

ax _____   cool _____   twig _____

ace _____   crunch _____   trim _____

ask _____   call _____   talk _____

answer _____   church _____   thin _____

▶ **Read the words at the bottom of the page. Cut them out and glue them in alphabetical order.**

Don't forget to look at the second letters if two words start with the same letter!

mask	test	dinner	much
dance	ears	eyes	taste

Name _____ Date _____

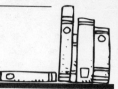

A Day at the Beach

➡ **Cut out the words. Glue them in alphabetical order on the lines.**

1. _____

2. _____

3. _____

4. _____

5. _____

6. _____

7. _____

8. _____

9. _____

10. _____

| summer | shovel | wind | pail |
| boat | tan | bucket | pour |

water

towel

© McGraw-Hill Children's Publishing   0-7424-1952-5 Complete Library Skills

Name _____ Date _____

Jumpin' Jellyfish

Jerry Jellyfish is all mixed up. The words below all start with the same letter.

▶ **Write these words in alphabetical order.**

grab   1. _____

ghost   2. _____

gust   3. _____

glue   4. _____

give   5. _____

▶ **Help Jerry number these words in alphabetical order.**

1. ☐peanut	2. ☐number	3. ☐blue	4. ☐sing
☐play	☐name	☐brush	☐strong
☐pop	☐none	☐bath	☐shy
☐paper	☐nine	☐bunch	☐swim
☐phone	☐never	☐bench	☐summer

© McGraw-Hill Children's Publishing   0-7424-1952-5 Complete Library Skills

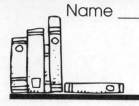 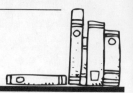

Name _____ Date _____

Who's Next?

▶ Look at each word below. Circle the word that would come next in the alphabet.

1. **game**	guest	glue	grip
2. **money**	mall	mug	Michigan
3. **spoon**	shrub	state	said
4. **pet**	prize	pizza	puppy
5. **ask**	all	after	ate
6. **ball**	bite	bait	better
7. **never**	nail	number	name
8. **cost**	came	crunch	chunk
9. **frost**	fast	fever	full
10. **hello**	happy	hippo	hungry

▶ On the bags below, write words of your own that could come after the word on the bag. Use a dictionary for help.

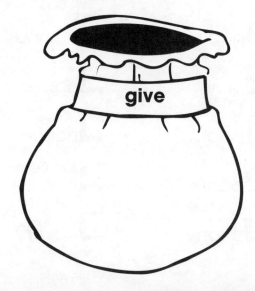

Name _____ Date _____

Nonfiction Books

Books that are about real things are called **nonfiction books**. They can be about dogs, airplanes, countries, planets, oceans, games, electricity—anything that is real. The title of a book can often tell you if the book is real or imaginary. The pictures in nonfiction books usually look real, too.

➤ **Look at the books below. Color the books that you think are nonfiction.**

➤ **Write the title and author of one of your favorite nonfiction books.**

_____

_____

Name _____ Date _____

Fiction Books

Fiction books are books that are about imaginary, make-believe things. They are not real. Fiction books can be about dogs that talk, people that fly, and magic carpet rides. Fiction books can take you to places you've only dreamed of going.

Like nonfiction books, fiction books can have titles that tell you that the book is make-believe.

▶ **Read the titles below. Color the books that you think are fiction.**

▶ **Write the title and author of one of your favorite fiction books.**

_____

_____

Name _____ Date _____

Biographies and Autobiographies

A **biography** is a book that is written about a person's life. It is written by another person. An **autobiography** is a book that a person writes about his or her own life.

▶ **Read the book titles below. If it sounds like a biography, color the leaf yellow. If it sounds like an autobiography, color the leaf green.**

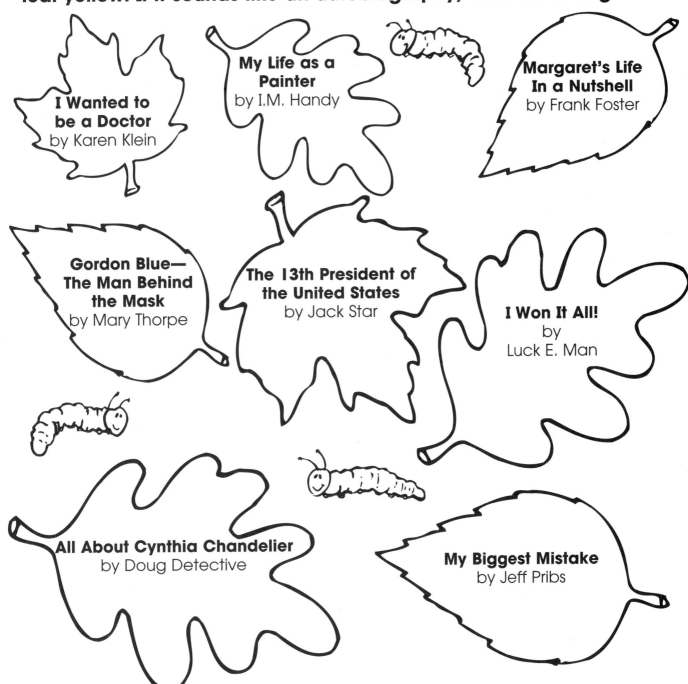

I Wanted to be a Doctor
by Karen Klein

My Life as a Painter
by I.M. Handy

Margaret's Life In a Nutshell
by Frank Foster

Gordon Blue—The Man Behind the Mask
by Mary Thorpe

The 13th President of the United States
by Jack Star

I Won It All!
by Luck E. Man

All About Cynthia Chandelier
by Doug Detective

My Biggest Mistake
by Jeff Pribs

Name _____ Date _____

Folktales and Fairy Tales

A **folktale** is a story that has been told over and over and passed down from generation to generation. A **fairy tale** is a type of folktale. Folktales usually begin with, "Once upon a time," or, "Long ago and far away." There are usually good and bad characters in the stories, and they often end, "happily ever after." Some well-known folktales are *Cinderella*, *Rupunzel*, and *Goldilocks and the Three Bears*.

▶ **Look at each book title below. Draw a line to connect it to the characters in the book.**

a.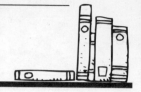

1. Little Red Riding Hood

2. Goldilocks and the Three Bears

c.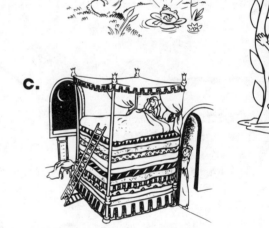

3. Jack and the Beanstalk

4. The Princess and the Pea

5. The Frog and the Princess

d.  e.

▶ **What is one of your favorite folktales or fairy tales?**

_____

Name _____ Date _____

Poetry

Poetry plays with language, words, and sounds. It often rhymes, but it doesn't have to. Poems can say a lot in only a few words or lines. They can be serious, funny, sad, mysterious, or just plain wacky.

▶ **Read the poem below. What kind of poem do you think it is?**

A fly and a flea in a flue
Were caught, so what could they do?
Said the fly, "Let us flee!"
"Let us fly!" said the flea,
And they flew through a flaw in the flue.

▶ **Write your own poem on the lines below.**

_____

_____

_____

_____

_____

Fiction vs. Nonfiction

To help reinforce the difference between fiction and nonfiction books, do the following activities with your students:

- Show your students books that will provide them with clear examples of the difference between fiction and nonfiction.

- Have a scavenger hunt in the library. Divide the class into two groups. Everyone in group one should choose a fiction book from the bookshelves. Everyone in group two should choose a nonfiction book. The group that has made the most accurate choices is the winner.

- Have your students make up titles of books. These titles should either reflect a fiction book or a nonfiction book.

- On a table in the center of your library, stack many fiction and nonfiction books. Let your students sort the books into two stacks—fiction and nonfiction. Or include books from all five genres covered on the last few pages and have your students sort them into five piles.

Name _____ Date _____

An Author's Name

When you see an author's name in the electronic card catalog, you will notice that the author's last name comes first. It is followed by the author's first name. These are separated by a comma.

▶ **Pretend you are an author. Write your name on the line, starting with your last name.**

_____

▶ **Look at the authors' names below. Rewrite them, starting with the last names.**

1. Ezra Jack Keats _____

2. Lois Lenski _____

3. Madeleine L'Engle _____

4. Mary Pope Osborne _____

5. J. K. Rowling _____

6. C. S. Lewis _____

7. Shel Silverstein _____

8. Louis Sachar _____

Name _____ Date _____

Fiction Call Numbers

Books are put on bookshelves in a certain order. They are arranged in alphabetical order according to their call numbers. Fiction call numbers begin with an *F* or *FIC* which stands for FICTION. Then you will see the first two letters of the author's last name. For example:

F Bo (or FIC Bo) is the call number for *My Dinosaur* **by Mark Bowen**

▶ **Look at the fiction book titles below. Write each call number on the line next to it. Don't forget to start it with an *F*.**

1. *The Lemonade Trick* by Scott Corbett _____
2. *The Real Me* by Betty Miles _____
3. *The Schoolhouse Mystery* by Gertrude C. Warner _____
4. *The Black Stallion* by Walter Farley _____
5. *Beezus and Ramona* by Beverly Cleary _____
6. *Sea Star* by Marguerite Henry _____
7. *Lone Hunt* by William O. Steele _____
8. *Homer Price* by Robert McCloskey _____
9. *Old Yeller* by Fred Gipson _____
10. *Across Five Aprils* by Irene Hunt _____

▶ **Arrange the above call numbers in alphabetical order.**

_____   _____
_____   _____
_____   _____
_____   _____
_____   _____

Name _____   Date _____

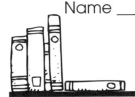 

Arrange the Fiction Books

▶ **Look at the fiction call numbers below. Cut them out and glue them in alphabetical order on the books. This is how you would find them on the bookshelves in the library.**

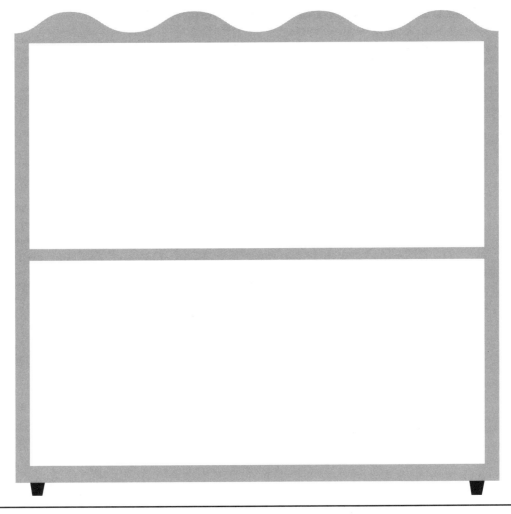

F	F	F	F	F	F	F	F	F	F
Qu	Br	Tu	Mo	An	Cr	No	Re	Or	Hi

Name _____ Date _____

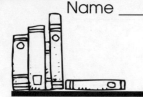

Shelving Fiction Books

The call numbers for fiction books are made up of the first two letters of the author's last name. These two letters are used to alphabetize the books on the bookshelves.

▶ **Look at the books below. Cut them out and glue them in alphabetical order on the bookshelf below.**

It's Like This, Cat
by
Emily Neville

Ginger Pye
by
Eleanor Estes

Rifles for Watie
by
Harold Keith

Up a Road Slowly
by
Irene Hunt

The Witch of Blackbird Pond
by
Elizabeth George Speare

Onion John
by
Joseph Krumgold

© McGraw-Hill Children's Publishing

0-7424-1952-5 Complete Library Skills

Name _____ Date _____

Nonfiction Call Numbers

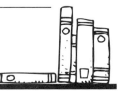

Like fiction call numbers, nonfiction call numbers show the first two letters of the author's last name. But nonfiction books have numbers that must be put in order *before* the books are alphabetized. These numbers are found on the spine of the book with the first two letters of the author's last name.

Libraries are divided into 10 nonfiction sections. These sections are part of the Dewey Decimal Classification® system. Each section has its own subject and numbers. For example, books on animals can all be found in the 500 section. The call number for *Why Bats Hang Upside Down* by Mike Wing would be: **546 Wi**

▶ **Look at the nonfiction call numbers below. You see a 3-digit number. You also see the first two letters of the author's last name. Write them in order, starting with the smallest number.**

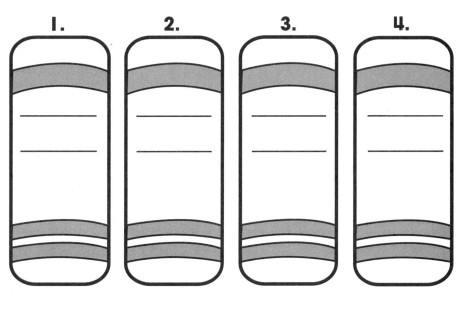

523	546	599	546
Ta	Br	We	Ba

▶ **Now look at the letters of the authors' last names. Are they in alphabetical order? If not, rearrange them so they are. Write them in the correct order on the lines below.**

_____ _____ _____ _____

© McGraw-Hill Children's Publishing   0-7424-1952-5 Complete Library Skills

Name _____ Date _____

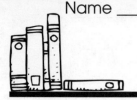

Shelving Nonfiction Books

To arrange nonfiction books on a bookshelf, you have to look at two things—the 3-digit number and the two letters of the author's last name. This is the call number.

▶ **Number these call numbers in alphabetical order using 1, 2, and 3.**

550   700   350
Zi   Ma   Cr
___   ___   ___

▶ **Look at the book spines below. Arrange them into the order you would find them in on a bookshelf. Write the call numbers in order on the bookshelf.**

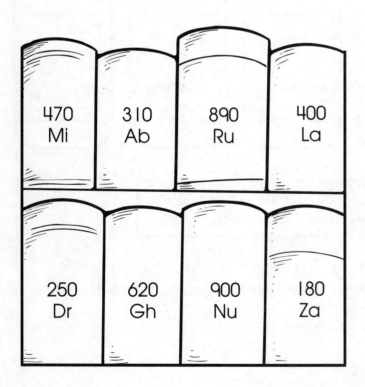

470 Mi   310 Ab   890 Ru   400 La
250 Dr   620 Gh   900 Nu   180 Za

© McGraw-Hill Children's Publishing

0-7424-1952-5 Complete Library Skills

Name _____ Date _____

Arranging Nonfiction Books

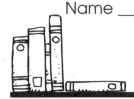

Nonfiction books that are in the same section in the library will all start with the same first number. Some books may even have the same three-digit number. When this happens, you must look at the first two letters in the author's last name and alphabetize them. The following call numbers are in alphabetical order:

        589   589   589
        Fa   Ma   St

➡ **The call numbers below all have the same 3-digit number. But they are not in alphabetical order. Write them in alphabetical order on the bookshelf.**

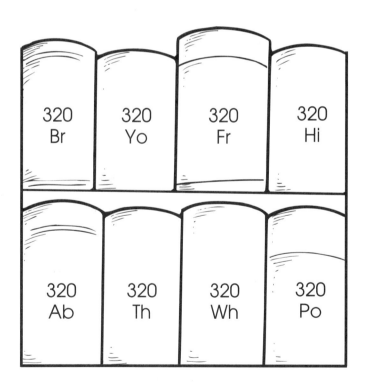

Name _____ Date _____

 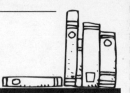

My Spine

▶ **Print the title of your favorite fiction or nonfiction book on the spine of this book. Then write the call number underneath it.**

Name _____ Date _____

The Ten Sections of the Library

The Dewey Decimal Classification® system—what a mouthful! Luckily it's easy to explain.

A long time ago, nonfiction books in the library were hard to find. They weren't arranged in any order. So a man named Melvil Dewey thought of a way to organize the thousands of books so they would be easier to find. He looked at what each book was about and put them into ten sections. Then he gave the sections numbers to help keep them apart. The Dewey Decimal Classification® system is below:

```
000-099—General Works (Encyclopedias)
100-199—Psychology, Philosophy
200-299—Religion
300-399—Fairy Tales, Folktales, Social Sciences
400-499—Languages
500-599—Sciences, Animals
600-699—Useful Arts (How Things Work)
700-799—Arts, Crafts, Sports
800-899—Poems, Plays, Short Stories
900-999—History, Geography, Biography
```

* Use this page as a reference for the next few pages.

Name _____ Date _____

The Dewey Decimal Classification® System

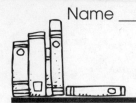

- Encyclopedias, Newspapers, Almanacs — 000-099
- Making Friends, Feelings, Optical Illusions — 100-199
- Bible Stories, Greek Myths, Amish, Judaism — 200-299
- Careers, Customs, Environment, Manners, Money — 300-399
- Sign Language, English, French, Chinese, Spanish, German, Italian — 400-499
- Animals, Biology, Plants, Dinosaurs, Insects, Rocks, Planets — 500-599
- Computers, Cookbooks, Human Body, Nutrition, Pets — 600-699
- Crafts, Games, Jokes, Music, Puppets, Sports — 700-799
- Writing, Plays, Poetry — 800-899
- Countries, Biographies, States, Travel, Wars — 900-999

© McGraw-Hill Children's Publishing   0-7424-1952-5 *Complete Library Skills*

Name _____ Date _____

Which Section Is It?

There are ten sections to the Dewey Decimal Classification® system.

▶ **Look at the books below. Write the Dewey Decimal Classification® system number on the front of each book. Use pages 95 and 96 for help.**

Our Country's Flag
1. _____

Indoor Pets
2. _____

Let's Travel to Ireland
3. _____

Cinderella
4. _____

French Dictionary
5. _____

My Poetry Book
6. _____

Kids' Recipes
7. _____

Brush and Canvas
8. _____

Tennis
9. _____

The Children's Bible
10. _____

Volcanoes and Earthquakes
11. _____

The T-Rex
12. _____

Name _____ Date _____

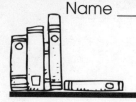 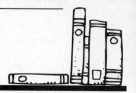

Dewey Match Game

Nonfiction books are divided into ten sections. This is called the Dewey Decimal Classification® system.

➡ **Read the book titles below. Draw a line to match each book with the section you would find it in. Use pages 95 and 96 for help.**

1. *Sign Language*
2. *Secrets of the Mind*
3. *All About Dogs*
4. *Best-Loved Poems*
5. *World War I*
6. *World Book Encyclopedia*
7. *Inside a Computer*
8. *Manners in a Public Place*
9. *Bible Stories*
10. *The Joke Book*

a. 000—Encyclopedias
b. 100—Psychology, Philosophy
c. 200—Religion
d. 300—Fairy Tales, Social Sciences
e. 400—Languages
f. 500—Sciences, Animals
g. 600—Useful Arts (How Things Work)
h. 700—Arts (Crafts, Games, Sports)
i. 800—Poems, Plays, Short Stories
j. 900—History, Geography, Biography

Name _____ Date _____

Dripping Dewey

The Dewey Decimal Classification® system is made up of ten sections.

▶ **Read the book subjects on the water drops below. Cut out the faucets and the water drops. On a separate sheet of paper, glue the drops under the sections (faucets) where they would be found in the library.**

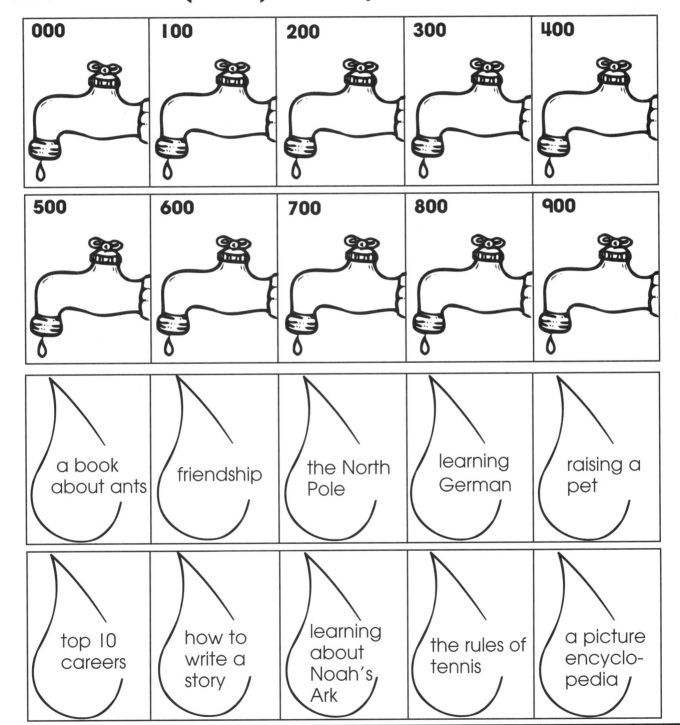

© McGraw-Hill Children's Publishing · 0-7424-1952-5 Complete Library Skills

 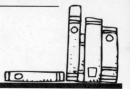

Name _____ Date _____

The Dictionary

The dictionary is a book of words. It shows you how to spell words and it tells you what words mean. It also tells you what part of speech a word can be, such as a noun (*n.*), verb (*v.*), or adjective (*adj.*).

▶ **Look at the following dictionary entry:**

bake (*n.*)
To cook food in an oven.
I like to **bake** *cookies for my mother.*

This dictionary entry shows you the word in bold. Then it tells you what the word means (the definition). Finally, the bold word is used in a sentence.

Some words can be used in more than one way in a sentence. For example:

cut
1. (*n.*) An opening in your skin that hurts.
   My **cut** *will not heal.*
2. (*v.*) To change the shape of something with scissors or a knife.
   My teacher **cut** *the paper in half.*

The word *cut* can be used as a noun and as a verb. The word is used both ways in two different sentences.

© McGraw-Hill Children's Publishing · 100 · 0-7424-1952-5 Complete Library Skills

Name _____ Date _____

Using a Dictionary

A dictionary tells you how to spell a word. It tells you what a word means. It also tells you how to use a word in a sentence. Dictionaries are arranged in alphabetical order. Some dictionaries also have pictures to go along with words.

➤ **Look at the dictionary page below. Answer the questions about it.**

Ff

face (n.)
The front part of your head.
*Your eyes, nose, and mouth are on your **face**.*

fall
1. (n.) A season of the year.
   ***Fall** is my favorite season of the year.*
2. (v.) To drop to the ground.
   *I saw my friend **fall** on the playground.*

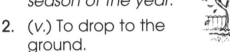

feathers (n.)
These keep birds warm.
*Sometimes I find **feathers** on the ground.*

fire (n.)
This happens when something burns.
*We like to build **fires** to keep warm.*

fly
1. (v.) To move through the air.
   *I wish I could **fly** like a bird.*
2. (n.) An insect with wings.
   *There is a **fly** in our house.*

frog (n.)
A small animal that lives near ponds and rivers.
*I like to watch **frogs** hop on lily pads.*

1. What part of speech is the word *frog*? _____
2. Which word is part of a bird? _____
3. The word *fall* can mean a _____ of the year.
4. Which words can be both a noun and a verb? _____
5. Write your own word that could be found on this page in a dictionary. _____

Name _____ Date _____

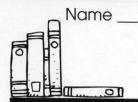
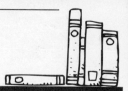

A-Maze-ing Spelling

▶ Betty Bee needs to take honey to her hive. Trace Betty's path by following the correctly spelled words below. Use a dictionary to check the spellings.

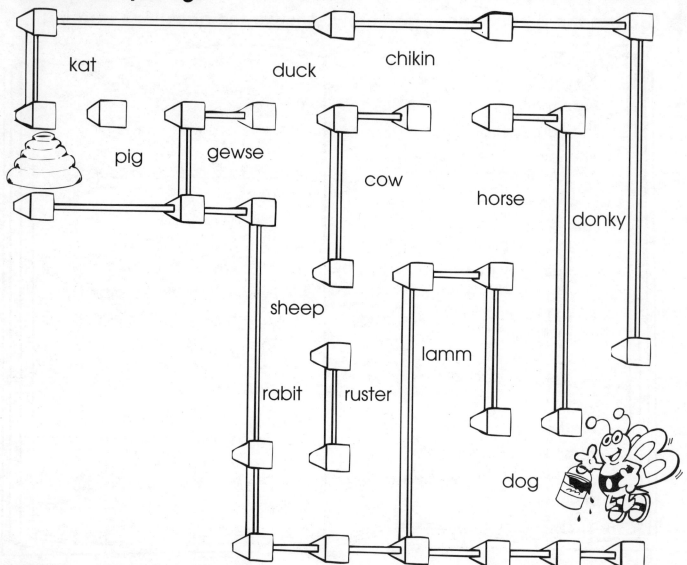

▶ **Bonus Bzzzy Work**

On the lines below, write the correct spellings of each misspelled word above. Use a dictionary.

_____ _____ _____

_____ _____ _____

Name _____ Date _____

Which Is Right?

➤ Use a dictionary to check the spellings of the words below. Write the correctly spelled words in the crossword puzzle.

Across
2. skait/skate
3. shert/shirt
6. cookie/cookey
8. airplain/airplane
9. only/onle

Down
1. sailor/sailer
4. howse/house
5. train/trane
6. caff/calf
7. knee/knea

Name _____   Date _____

The Encyclopedia

An **encyclopedia** is a set of books. Each book is called a **volume**. People use encyclopedias to learn about many different things. They are arranged in alphabetical order on bookshelves.

Using an encyclopedia, you can learn about famous people, faraway countries, the solar system, the brain, and so much more. This is a good place to look if you need to learn about something for a report.

Each volume has a letter or two on its spine. These letters tell you what subjects you will find inside this volume. For example:

> If you are doing a report on Mexico, you want to look for the volume that has an *M* on its spine. This tells you that subjects starting with the letter *M* are inside this volume. What else do you think you could find inside Volume M? List a few subjects that start with the letter *M*.

_____   _____

_____   _____

▶ **You can do a report on anything in Volume ____. Look in this volume and write down subjects that you could do your report on.**

A	B	C-Ch	Ci-Cz	D	E	F	G	H	I
1	2	3	4	5	6	7	8	9	10

J-K	L	M	N	O-P	Q-R	S	T	U-V	W-X Y-Z
11	12	13	14	15	16	17	18	19	20

_____

_____

_____

_____

_____

_____

Name _____ Date _____

Using an Encyclopedia

Encyclopedias are filled with information. You can learn about spider monkeys, China, the human body, and so much more. But you have to know where to look.

Each encyclopedia has its own letter. A few have more than one letter. These letters are on the spines. If the subject you want to learn about starts with a *t*, such as toys, look in Volume T.

▶ **Look at the subjects below. Draw a line to connect each subject to the volume you would find it in.**

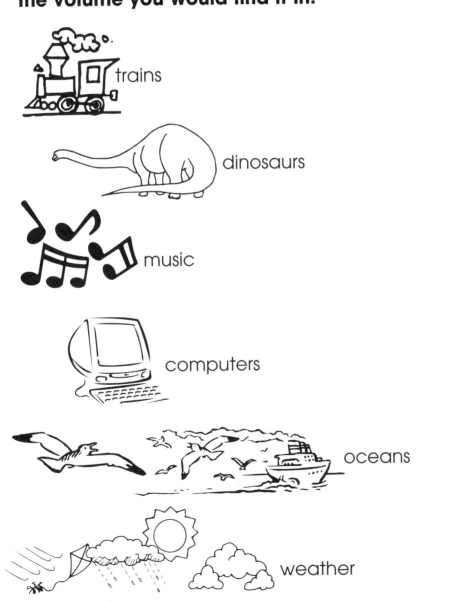
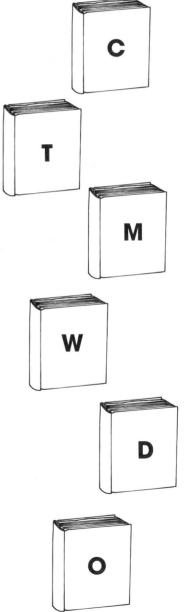

- trains
- dinosaurs
- music
- computers
- oceans
- weather

C
T
M
W
D
O

Name _____ Date _____

Dictionary and Encyclopedia Review

➤ Read each section of the centipede. If you can use a dictionary or encyclopedia to find the information in the section, color the section red. If you have to use another book, color it black.

- How to say the word *stitch*
- Kinds of turtles
- Directions for finding a buried treasure
- Definition for the word *teepee*
- A joke to tell your friends
- Types of music
- Your principal's favorite color
- How to use the word *fly* in a sentence
- A recipe for chicken noodle soup

Name _____ Date _____

Guide Words

When you open a dictionary, you will see a word at the top of every page. These words are called **guide words**. They help guide you to the word you are looking for. Look at the dictionary pages. The word *bake* is the first word on the left page. The word *bug* is the last word on the right page. Any words that come between *bake* and *bug* can be found on these two pages.

▶ **Can you think of words that might be on these pages? Write them on the lines.**

_____   _____

_____   _____

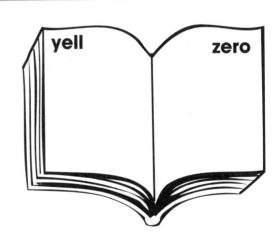

▶ **Look at these dictionary pages. The first word on the left page is _____. The last word on the right page is _____.**

▶ **What words might be found on these pages?**

_____   _____

_____   _____

Name _____ Date _____

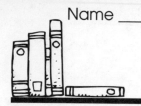
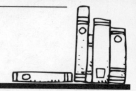

Using Guide Words

Dictionary pages have words that help you find what you are looking for. They are called **guide words**. You will see guide words at the top of every page.

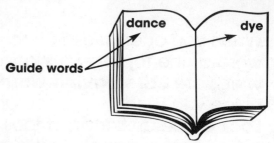

▶ Match each word in the Word Bank to the dictionary page on which it can be found. Use the guide words for help. Write each word on the correct book.

Word Bank

| out | answer | drag | have | pot |
| knew | clown | stop | vapor | |

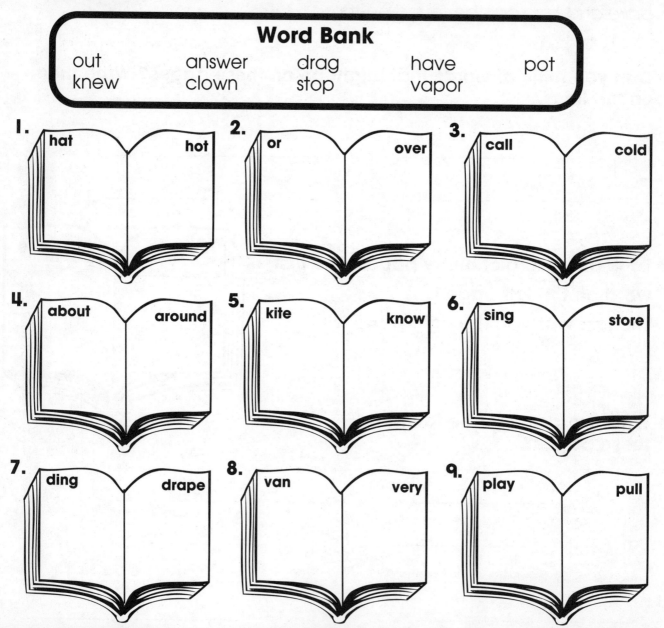

1. hat — hot
2. or — over
3. call — cold
4. about — around
5. kite — know
6. sing — store
7. ding — drape
8. van — very
9. play — pull

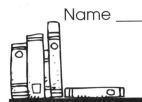

Name _____ Date _____

Guide Me

Guide words help you find words in a dictionary. You will see guide words at the top of every page. Some dictionaries have two guide words on every page. The first guide word is the first word on that page. The second guide word is the last word on that page.

➤ **Look at the guide words below. Write each word from the Word Bank on the page where it would be found in the dictionary. If a word does not belong on these pages, write it below the book.**

Word Bank

man	mask	hat	lost	horse	green	make
lip	love	most	noon	girl	happy	grip

ghost   house   limb   mat

Name _____ Date _____

Find the Guide Words

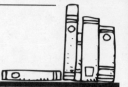

Guide words can be very helpful. They can guide you to the word you are looking for in a dictionary.

➤ **Look up the words below in a dictionary. Write the two guide words that are on the top of the page.**

1. mouse _____ _____
2. down _____ _____
3. bingo _____ _____
4. sail _____ _____
5. police _____ _____
6. attract _____ _____
7. vermin _____ _____
8. clover _____ _____
9. petal _____ _____
10. wrist _____ _____
11. talent _____ _____
12. danger _____ _____

Name _____ Date _____

Finding a Word

Dictionaries can be divided into three sections—the beginning, middle, and end.

In the beginning, you will find words that start with A–H. In the middle, you will find words that start with I–Q. The end is where you will find words that start with R–Z.

Remember this to help you find words in a dictionary.

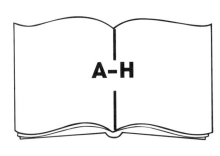
A–H

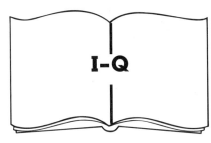
I–Q

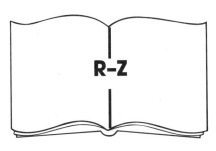
R–Z

▶ **Use the dictionaries above to help you answer the questions below. Write beginning, middle, or end on the line. In a dictionary, where would you find the word ...**

1. basket? _____
2. play? _____
3. magnet? _____
4. zoo? _____
5. color? _____
6. windy? _____
7. flower? _____
8. quack? _____
9. rose? _____
10. huge? _____

Name _____ Date _____

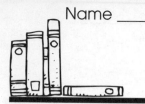

Which Section Is It In?

A dictionary is a big book with a lot of words in it. Where do you even start looking? First divide the dictionary up into three sections in your mind—the beginning, middle, and end. Now it doesn't seem so big.

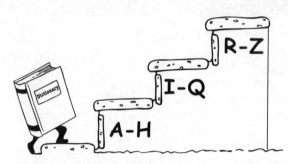

▶ **Cut out the words below. Look at the first letter in each word. Decide where the word can be found in a dictionary. Glue it in the correct section.**

Beginning	Middle	End

why	after	read	just
sleep	those	dear	idea
answer	limb	open	hungry

© McGraw-Hill Children's Publishing   0-7424-1952-5 Complete Library Skills

Name _____ Date _____

Dictionary Dividers

The dictionary has three sections—A–H, I–Q, and R–Z.

➡ **Fill in the missing letters of the alphabet. Then show which letters the words would come between in a dictionary.**

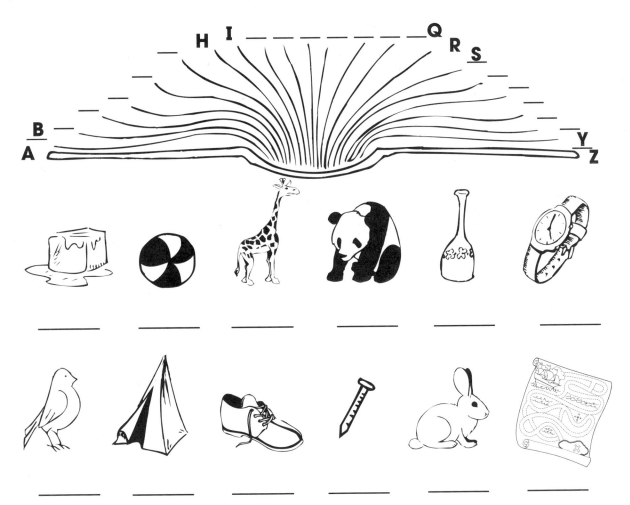

Write the words you found in the middle of the dictionary (I–Q). Circle the last letter in each word. Then unscramble the letters.

_____

What do you call a group of leopards?

Answer: _____

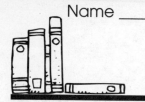

Name _____ Date _____

Divide the Dictionary

To find words easily in a dictionary, divide it into three sections— A–H, I–Q, and R–Z. Look at the first letter in a word to decide which section it would be in.

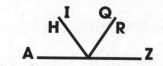

▶ Write each word from the Word Bank in the section of the dictionary where you would find it. Write them on the lines below.

Word Bank

whale	juice	geese	always	door	lizard
queen	music	tree	keep	empty	nobody
coffee	brown	young	volume	unfair	rainbow

Beginning   **Middle**   **End**

_____   _____   _____
_____   _____   _____
_____   _____   _____
_____   _____   _____
_____   _____   _____
_____   _____   _____

▶ Choose one word from each section. Find each one in a dictionary. Write the page number and the words that come before and after it.

| Word | Page | Before and After |

1. _____ _____ _____ and _____
2. _____ _____ _____ and _____
3. _____ _____ _____ and _____

Name _____ Date _____

Round 'Em Up!

▶ Look at the words in the Word Bank. Decide where each word would be found in a dictionary. Search inside each lasso for the words. Circle the words.

BEGINNING A–H

END R–Z

Word Bank
weak
tassel
awning
baby
finger
veto
aid
sip
glow
ranch
all
tint
dim
full
uncle

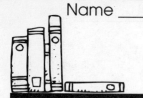
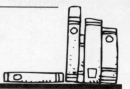

Lasso Lily

▶ Look at the words in the Word Bank. Decide where each word would be found in a dictionary. Search inside each lasso for the words. Circle the words.

Word Bank

net
jail
dock
old
vase
inn
zone
high
many
bed
tow
saw
fist
brick
tan
go
kite
race
cat
yes
well
pan
lamp

BEGINNING A–H

K H T B E D
T G O K T O
F L F H A C
I B R I C K
S H I G H C
T N G P S T

MIDDLE I–Q

M N E T T J
L W J I N N
A M A N Y E
M R I S N T
P O L D L I
S I P A N K

END R–Z

W Z A E V T
A V R A C E
T A N O S W
S S Z O N E
A E S Z P L
W Y T O W L

Name _____ Date _____

I'm Divided

▶ You have learned that dictionaries can be divided into three sections. What are those sections? Write them in the diagram.

▶ Using a dictionary, find ten words and write them below. Next to each word, write the section of the dictionary that you found it in.

Word	Dictionary Section
1. _____	_____
2. _____	_____
3. _____	_____
4. _____	_____
5. _____	_____
6. _____	_____
7. _____	_____
8. _____	_____
9. _____	_____
10. _____	_____

Name _____ Date _____

Know Your Computer

Before you use a computer, you should know what the different parts are. Each part has its own job.

▶ **Look at the parts of a computer below. Draw a line to match the part of the computer on the left to its job on the right.**

Keyboard
•

• A small part that is attached to the computer that lets you move around on the screen.

Mouse
•

• The face of the computer. This is where you see what you have written or drawn.

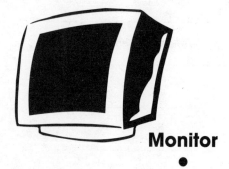
Monitor
•

• This is the brain of the computer. It does what you tell it to do.

CPU
•

• This has all of the letters, numbers, and commands that you need.

Name _____ Date _____

Keyboard Practice

Learning to use the keyboard is very important. Practicing over and over again is the best way to become better at typing.

▶ **Follow the directions below to practice typing and making sentences.**

1. Open your word processing program.

2. Type the following words: **dog   frog   cat**

3. Use the [Delete] or [Backspace] key to delete the word **cat**.

4. Type the following sentence: **My bike is red.**
   Use the [Shift] to make the **m** a capital letter.

5. Use the [Delete] or [Backspace] key to delete the word **red**.

6. Now, type the word **blue** so the sentence is: **My bike is blue.**

7. Type a sentence of your own.

8. Quit or exit the program.

Name _____ Date _____

You're an Artist

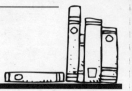

One thing you can do on a computer is draw. You can create new pictures, paint a picture with color, stamp letters, numbers, and objects on the screen, and much, much more.

▶ **Follow the directions below to practice drawing on a computer.**

1. Open your drawing program.

2. Click on the [✏️]. Draw a straight line. _____

   Draw a curved line. ∿∿∿

3. Click on the [🧽]. Drag the eraser over your lines and see what happens.

4. Click on the [A]. Type your name on the screen.

5. Click on the [🔨]. Stamp a pattern using two shapes.

6. Click on the [✏️] again. Draw a full circle.

7. Click on the [🪣] to color your circle. Choose a color. Paint your circle.

▶ **Continue clicking on buttons to see what you can do with them. The best way to learn about your drawing program is to experiment and have fun with it.**

A Nonfiction Book Project

Show your students how to use the resources in the library to learn about things that interest them. If they know that their school library is filled with fun, interesting books, they will be anxiously awaiting every trip to the library.

Have your students choose an animal that interests them. Using nonfiction books in the library, have your students research their animal. Encourage them to choose more than one book to get a variety of facts and to see many pictures that will help them better understand the animal they've chosen. Tell your students that they will use the information they find to create an animal report that they will share with the class.

Before your students start their research, explain to them the importance of putting information from a book into their own words. Use a book and an overhead to demonstrate how to record information using only a few words or a short sentence. As a class, form a paragraph on the overhead using the notes you have made. This should help them choose important words about their animals that they can later use in their animal reports. Students can record their notes on page 122.

My Stuffed Animal

After your students have finished their research, have them create a stuffed version of the animal they chose. They should use their facts to ensure that their stuffed animal is realistic. Follow the steps below for making a stuffed animal:

1. Sketch your animal on a large piece of paper. Include details, such as facial features and markings.
2. Cut out two copies of your sketch. The decorated side will be the front of your stuffed animal. The other side will be the back.
3. Color, paint, and decorate your animal to make it look realistic.
4. Staple the lower edge and two sides of your animal so that it is half closed. Stuff shredded newspaper or paper towels into the pouch you have created. Use enough paper so that it fills out, but don't overstuff it. Now staple the rest of the animal closed. Your stuffed animal is ready to show off!

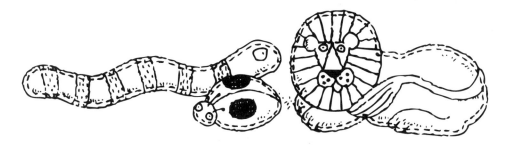

Name _____ Date _____

My Animal Report Notes

My animal is a _____.

Describe the animal (include its size, coloring, body markings, etc.).
_____
_____

What is its habitat and why does it live there?
_____
_____
_____

What does it eat?
_____
_____
_____

Does it have enemies? What are they?
_____
_____
_____

List one important fact about this animal.
_____
_____
_____

Name _____ Date _____

How Did I Do?

➤ **Color the face that shows how you did.**

	:)	:|	:(
I used more than one book for my research.	☺	😐	☹
I took notes in my own words.	☺	😐	☹
I knew where to look in the library for my nonfiction books.	☺	😐	☹
I knew how to use the table of contents in the books to find what I was looking for.	☺	😐	☹
I checked to make sure my spelling was correct.	☺	😐	☹
I gave the report my best effort.	☺	😐	☹

Besides the nonfiction books that you used, where else could you have found information about your animal?

_____

What did you enjoy about doing your animal report?

_____

_____

© McGraw-Hill Children's Publishing 0-7424-1952-5 Complete Library Skills

Name _____ Date _____

Library Awards/Reminders

Through rain or shine, I'm a WHIZ on the computer!

There's no way around it, is hooked on books!

I know how to use the library.

Look Mom, no hands!

Newbery and Caldecott Book Award Winners

The following lists highlight the last 15 years of Newbery and Caldecott award winners.

Newbery Award Winners

2003: *Crispin: The Cross of Lead* by Avi
2002: *A Single Shard* by Park, Linda Sue
2001: *A Year Down Yonder* by Peck, Richard
2000: *Bud, Not Buddy* by Curtis, Christopher Paul
1999: *Holes* by Sachar, Louis
1998: *Out of the Dust* by Hesse, Karen
1997: *The View from Saturday* by Konigsburg, E. L.
1996: *The Midwife's Apprentice* by Cushman, Karen
1995: *Walk Two Moons* by Creech, Sharon
1994: *The Giver* by Lowry, Lois
1993: *Missing May* by Rylant, Cynthia
1992: *Shiloh* by Naylor, Phyllis Reynolds
1991: *Maniac Magee* by Spinelli, Jerry
1990: *Number the Stars* by Lowry, Lois
1989: *Joyful Noise: Poems for Two Voices* by Fleischman, Paul

Caldecott Award Winners

2003: *My Friend Rabbit* by Rohmann, Eric
2002: *The Three Pigs* by Wiesner, David
2001: *So You Want to Be President?* by Small, David
2000: *Joseph Had a Little Overcoat* by Taback, Simms
1999: *Snowflake Bentley* by Azarian, Mary
1998: *Rapunzel* by Zelinsky, Paul O.
1997: *Golem* by Wisniewski, David
1996: *Officer Buckle and Gloria* by Rathmann, Peggy
1995: *Smoky Night* by Diaz, David
1994: *Grandfather's Journey* by Say, Allen
1993: *Mirette on the High Wire* by McCully, Emily Arnold
1992: *Tuesday* by Wiesner, David
1991: *Black and White* by Macaulay, David
1990: *Lon Po Po: A Red-Riding Hood Story from China* by Young, Ed
1989: *Song and Dance Man* by Gammell, Stephen

Suggested Authors for K–2

The following is only a short list of the many books available for young students. Libraries and the Internet are good sources for more books that are appropriate for the K–2 grade level.

Kindergarten

Carlo Likes Reading by Spanyol, Jessica
Chicka Chicka Boom Boom by Archambault, John
Corduroy by Freeman, Don
Curious George by Rey, H. A.
Little Miss Spider by Kirk, David
Madeline by Bemelmans, Ludwig
Stone Soup by Brown, Marcia
The Giving Tree by Silverstein, Shel
The Paper Bag Princess by Munsch, Robert
The Rainbow Fish by Pfister, Marcus
The Very Hungry Caterpillar by Carle, Eric
The War Between the Vowels and the Consonants by Turner, Priscilla
Where the Wild Things Are by Sendak, Maurice
Wemberly Worried by Henkes, Kevin

First Grade

A Birthday for Frances by Hoban, Russell
Angelina Ballerina by Holabird, Katharine
Annie and the Wild Animals by Brett, Jan
Arthur's Underwear by Brown, Marc
Good Night, Good Knight by Thomas, Shelley Moore
Harry in Trouble by Porte, Barbara Ann
Henry and Mudge and the Happy Cat by Rylant, Cynthia
ABC I Like Me! Carlson, Nancy L.
Imogene's Antlers by Small, David
Katy and the Big Snow by Burton, Virginia Lee
Little Bear's Friend by Minarik, Else Holmelund
Marvin One Too Many by Paterson, Katherine
Mouse Soup by Lobel, Arnold
Mouse Tales by Lobel, Arnold
Owl at Home by Lobel, Arnold
Ringo Saves the Day by Clements, Andrew
The Grouchy Ladybug by Carle, Eric
The Patchwork Quilt by Flournoy, Valerie
The Wednesday Surprise by Bunting, Eve

Second Grade

A Chair for My Mother by Williams, Vera B.
A Tree is a Plant by Bulla, Clyde Robert
Abigail Takes the Wheel by Avi
Abiyoyo by Seeger, Pete
Adventures of Taxi Dog by Barracca, Deborah
Amelia Bedelia by Parish, Peggy
Arthur's Mystery Envelope by Brown, Marc
Ben Franklin and His First Kite by Krensky, Stephen
Breakout at the Bug Lab by Horowitz, Ruth
Cloudy With a Chance of Meatballs by Barrett, Judi
Henry and Mudge: The First Book of Their Adventures by Rylant, Cynthia
Junie B., First Grader: Boss of Lunch by Park, Barbara
Keep the Lights Burning, Abbie by Roop, Connie
Magic Tree House Series by Osborne, Mary Pope
Mr. Putter and Tabby Pick the Pears by Rylant, Cynthia
Mrs. Piggle-Wiggle by Macdonald, Betty
Pinky and Rex by Howe, James
Sylvester and the Magic Pebble by Steig, William
Tacky the Penguin by Lester, Helen
The Drinking Gourd by Monjo, F. N.
The Mitten by Brett, Jan
The Velveteen Rabbit by Williams, Margery

Answer Key

Library Citizenship 14
Quiet Cara
not talking
sitting still
raising hand
gently holding a book
Noisy Nathan
talking
moving all over
interrupting
throwing a book

Cara is showing good library behavior.

Alphabetizing by Author's Last Name 23
Swish Swish by Tim <u>B</u>ubble
Ice by Amy <u>F</u>reeze
Sticky Stuff by Johnny <u>J</u>elly
Trot Along by Neil <u>N</u>ay
Puppies by Judy <u>R</u>uff
Apples by Jim <u>S</u>tem

Missing Books 24
Missing letters: B, C, E, G, I, J, L, N, O, Q, R, U, V, X, Z

Keys to Reading Success 32
Red spaces:
Wash hands before touching books
Carry books home in a bag
Return my book on time
Use a bookmark

Other spaces should be colored black.

Putting Books on a Shelf 35
1. marker   2. out
3. gently   4. book
5. tall

Finding the Author 39
Star Myers
Jerry Geranium
Shiny Armor

Alphabet Fun 44
Missing letters: c, d, f, h, i, j, k, l, m, o, q, r, s, u, v, x y, z
Answer to the riddle: your shadow

In the Library 46
1. author   2. books
3. computer   4. illustrator
5. librarian   6. quiet
7. read   8. stories

ABCs to the Rescue 47
Max should follow these words: age, bug, chip, door, ears, four, and grow.
Max found his puppy swimming in the pond.

The Alphabet Divided 48
A–H: ate, food, help
I–Q: jump, keep, look
R–Z: ripe, say, vase

The Title Page 51
title: *My Dog*; author: Jill Woof

The Table of Contents 52
cats—page 5; bats—page 3; bees—page 7

The Table of Contents 53
fish—page 7; frogs—page 3; bugs—page 5

The Table of Contents 54
comets—page 6; moon—page 9; sun—page 12; page 3—stars; page 15—planets

The Table of Contents 55
page 6—whales; page 12—seals; sharks—page 3; penguins—page 15; octopuses—page 9

Fiction and Nonfiction Books 58
fiction book: *The Mouse and the Talking Cheese*; nonfiction book: *Taking Your Puppy to the Vet*

Fiction or Nonfiction? 59
fiction books: *The Story of Brown Bear*, *The Little Tree That Could Sing*, *The Flying Car*

Fiction or Nonfiction? 60
fiction books: *Three Naughty Mice*, *The Magical Web*, *Sam, the Purple Frog*

Fiction or Nonfiction? 61
nonfiction books: *The Fish Book*, *All About Insects*, *The Dog Book*

Call Numbers 62
1. Al   2. Fl

Blast Off! 70
Yellow spaces: return books on time, whisper, replace books correctly on shelves, take turns, use a shelf marker; all other spaces should be red.

The Title Page and The Copyright Page 73
1. *Raising a Puppy*
2. Jeff Ruff
3. New York
4. Wag Publishers
5. no
6. no
7. yes

The Table of Contents 74
1. page 18
2. Chapter 4, page 12
3. Buzzing Bees
4. When to Pick Vegetables
5. page 6
6. 9 chapters
7. Chapter 8
8. about 30 pages
9. no
10. Answers will vary.

The Parts of a Book 75
1. title
2. table of contents
3. author
4. copyright page
5. spine
6. illustrator
7. title page
8. cover
9. publisher

Which Book Comes Next? 76
Ad, Dh, Hl, Mq

Wait a Second! 77
ace, answer, ask, ax
call, church, cool, crunch
talk, thin, trim, twig
dance, dinner, ears, eyes,
mask, much, taste, test

A Day at the Beach 78
boat, bucket, pail, pour, shovel, summer, tan, towel, water, wind

Jumpin' Jellyfish 79
ghost, give, glue, grab, gust
1. paper, peanut, phone, play, pop
2. name, never, nine, none, number
3. bath, bench, blue, brush, bunch
4. shy, sing, strong, summer, swim

Answer Key

Who's Next? 80
1. glue   2. mug
3. state   4. pizza
5. ate   6. better
7. number   8. crunch
9. full   10. hippo

Nonfiction Books 81
nonfiction books: *The Storm of 2004, Space Travel, Riding on a Dogsled*

Fiction Books 82
fiction books: *Roger Goes to the Beach, Attack of the Gummy Monkeys, I'm an 8-Year-Old Spy*

Biographies and Autobiographies 83
yellow leaves: *The 13th President of the United States, Gordon Blue— The Man Behind the Mask, Margaret's Life in a Nutshell, All About Cynthia Chandelier*
green leaves: *My Life as a Painter, I Won It All!, My Biggest Mistake, I Wanted to be a Doctor*

Folktales and Fairy Tales 84
1. e   2. d   3. b   4. c
5. a

An Author's Name 87
Keats, Ezra Jack; Lenski, Lois; L'Engle, Madeleine; Osborne, Mary Pope; Rowling, J. K.; Lewis, C. S.; Silverstein, Shel; Sachar, Louis

Fiction Call Numbers 88
Alphabetical order: Cl, Co, Fa, Gi, He, Hu, Mc, Mi, St, Wa

Arrange the Fiction Books 89
An, Br, Cr, Hi, Mo, No, Or, Qu, Re, Tu

Shelving Fiction Books 90
Ginger Pye, Up a Road Slowly, Rifles for Watie, Onion John, It's Like This, Cat, The Witch of Blackbird Pond

Nonfiction Call Numbers 91
523 Ta, 546 Ba, 546 Br, 599 We

Shelving Nonfiction Books 92
350 Cr, 550 Zi, 700 Ma
180 Za, 250 Dr, 310 Ab, 400 La, 470 Mi, 620 Gh, 890 Ru, 900 Nu

Arranging Nonfiction Books 93
320 Ab, 320 Br, 320 Fr, 320 Hi, 320 Po, 320 Th, 320 Wh, 320 Yo

Which Section Is It? 97
1. 900–999   2. 500–599
3. 900–999   4. 300–399
5. 400–499   6. 800–899
7. 600–699   8. 700–799
9. 700–799   10. 200–299
11. 500–599   12. 500–599

Dewey Match Game 98
1. e   2. b   3. f
4. i   5. j   6. a
7. g   8. d   9. c   10. h

Dripping Dewey 99
000—a picture encyclopedia
100—friendship
200—learning about Noah's ark
300—top 10 careers
400—learning German
500—a book about ants
600—raising a pet
700—the rules of tennis
800—how to write a story
900—the North Pole

Using a Dictionary 101
1. noun
2. feathers
3. time or season
4. fall and fly

A-Maze-ing Spelling 102
Correctly spelled words: dog, horse, cow, sheep, duck, pig

Which Is Right? 103
Across   Down
2. skate   1. sailor
3. shirt   4. house
6. cookie   5. train
8. airplane   6. calf
9. only   7. knee

Using an Encyclopedia 105
trains—T; dinosaurs—D; music—M; computers—C; oceans—O; weather—W

Dictionary and Encyclopedia Review 106
Red sections: How to say the word *stitch*, Kinds of turtles, Definition for the word *teepee*, Types of music, How to use the word *fly* in a sentence.

Other spaces should be colored black.

Using Guide Words 108
1. have   2. out
3. clown   4. answer
5. knew   6. stop
7. drag   8. vapor
9. pot

Guide Me 109
ghost–house—hat, horse, girl, green, happy grip
limb–mat—man, lip, mask, love, lost, make
Remaining words: most, noon

Finding a Word 111
1. beginning   2. middle
3. middle   4. end
5. beginning   6. end
7. beginning   8. middle
9. end   10. beginning

Which Section Is It In? 112
Beginning words: after, dear, answer, hungry
Middle words: just, idea, limb, open
End words: why, read, sleep, those

Dictionary Dividers 113
ice—I-Q; ball—A-H; giraffe—A-H; panda—I-Q; vase—R-Z; watch—R-Z; bird—A-H; tent—R-Z; shoe—R-Z; nail—I-Q; rabbit—R-Z; map—I-Q
Answer: leap

Divide the Dictionary 114
Beginning: coffee, brown, geese, always, door, empty
Middle: queen, juice, music, keep, lizard, nobody
End: whale, tree, young, volume, unfair, rainbow

Round 'Em Up! 115
Beginning: awning, baby, finger, aid, glow, all, dim, full
End: weak, tassel, veto, sip, ranch, tint, uncle

Lasso Lily 116
Beginning: dock, high, bed, fist, brick, go, cat
Middle: net, jail, old, inn, many, kite, pan, lamp
End: vase, zone, tow, saw, tan, race, yes, well